'[Sue Halpern] writes as a journalist of the new, reflective, kind who does science, and sometimes literature, a service by travelling into the field with its practitioners and making both them and their work come alive . . . Sue Halpern's fluent, questing, and action-packed narrative presents a memorable gallery of biologists, but also attends to the army of amateurs whose observations of monarchs across a continent . . . are helping to fill in the map of evidence'   Michael Viney, *Irish Times*

'The magic of this book is that through Halpern's narrative, we witness her own unfolding obsession, witness that she is also "caught in the mystery of the Monarch Butterfly", drawn to an insect whose behaviour seems "so unlikely, so amazing, so nearly heroic". The quest truly does become her own, and we are taken on a journey filled with wonder, delight and passion, a first hand account of newly found excitement and intrigue'
Romy Needham, *Yorkshire Post*

'This delightful book . . . I learned much about the human dimensions of the 50-year unfolding drama of the North American monarch . . . This outstanding piece of narrative sociology is essential reading for monarch-fancier fanciers – and anyone curious about how natural history really gets done'   Richard Vane-Wright, *New Statesman*

Sue Halpern was educated at the universities of Yale and Oxford. She has written for the *New York Times*, *Granta* and the *New York Review of Books*, and her last book *Migrations to Solitude* was a *New York Times* Notable Book of the Year. She lives with her husband and daughter in a small town in the Adirondack Mountains of New York, where she helped start the first public library.

By Sue Halpern

Migrations to Solitude
Four Wings and a Prayer

Four Wings and a Prayer

Caught in the Mystery of the Monarch Butterfly

SUE HALPERN

PHOENIX

A PHOENIX PAPERBACK

First published in Great Britain in 2001
by Weidenfeld & Nicolson
This paperback edition published in 2002
by Phoenix,
an imprint of Orion Books Ltd,
Orion House, 5 Upper St Martin's Lane,
London WC2H 9EA

This edition is published in association with Pantheon Books,
a division of Random House, Inc., New York.

A CIP catalogue record for this book
is available from the British Library.

ISBN 0 75381 336 X

Printed and bound in Great Britain by
The Guernsey Press Co. Ltd, Guernsey, C.I.

To Sophie Crane McKibben
and her devoted scout, Barley,
who brought me outside

FOUR WINGS AND A PRAYER

Chapter 1

BILL CALVERT EASED his truck off Interstate 281 near McAllen, Texas, pulled into a mall parking lot, and drew a knife from his knapsack. It was late in the day, about eight o'clock, and he had been driving for the past five hours.

"What you want to do is make the cut like this," he said, unfastening his belt buckle and the top button of his jeans. He peeled back the waistband to reveal the smallest of incisions. "Nothing too obvious."

Calvert pressed on the fabric, and it opened, exposing a tunnel the width of two fingers. He reached in and extracted a wad of cash that was folded to the size and shape of a stick of gum. Three hundred dollars, it looked like.

"You try," he said, handing over the knife.

I got out of the truck and began to slice at the inside of my

3

jeans. People walked by, mothers and fathers towing small children, for the most part, but also the occasional solitary individual or couple, and if they found it odd to see a woman with a knife in her hand fiddling with her pants not two hundred yards from Montgomery Ward, they weren't saying.

"It's so uncomfortable to walk around with money in your shoes," Calvert was explaining. "It gets real damp. And smelly. This is much better."

We were ten miles from the Mexican border. I threaded my money into its hideaway and followed Calvert into the mall restaurant, a Luby's cafeteria. We were the only diners.

"I always come here before I go to Mexico," he said happily, sliding his tray along the steam table and overloading it with plates of green beans, broccoli, and peas, all of which looked like they had been through the wash. "These are the last green vegetables we'll see for two weeks." I couldn't say I was sorry.

BILL CALVERT is a biologist. Not the kind of biologist who wears a lab coat and not, especially, the kind who has a lab. He works out-of-doors most of the time, observing and cataloging and trying to come to terms with natural phenomena. Among people who study monarch butterflies, which is what he himself has done for the past twenty-five years, Calvert is considered the best field researcher in the pack. This may have something to do with the fact that Bill Calvert isn't really *part* of a pack. He works by himself, getting grants here and there and leading trips for science teachers and wealthy ecotourists, just scraping by. Although he has a doctorate in zoology, academia doesn't interest him. A "real" job doesn't interest him. Calvert is fifty-eight years old. Going to Mexico to look for monarchs—what we were doing—interests him.

"I TOOK AN aptitude test when I was in my thirties, and I scored two sigmas past a seventeen-year-old for 'desire for adventure,'" he said a couple of hours after we crossed the border at Reynosa, as he leapfrogged eighteen-wheelers along the rutted two-lane Mexican highway to Ciudad Victoria, where we planned to stop. It was close to midnight. We had just breezed through two military checkpoints with the words "*Biologico*" and "*Mariposa monarca*," and now ours was the only passenger vehicle on the road.

"I thought you weren't supposed to drive at night in Mexico," I said to Bill, who smiled at me and tugged on his mustache.

"Why not?" he asked thoughtfully, as if it were a real question.

"Bandits," I said.

"Maybe," he said, and smiled again. It was an enigmatic smile, nothing comforting.

I DIDN'T KNOW Bill Calvert. Or rather, I had known Bill Calvert for about ten hours, ever since he picked me up at the Austin airport earlier in the day. He was late, and I had begun to have my doubts, but then he'd rushed through the door and though I had never before laid eyes on him, I recognized him instantly and was reassured. He was a familiar *type*. Tall, thin, with a professorial mop of graying hair and an abundant white mustache, wearing stiff Wranglers, a plaid shirt, and scuffed brown shoes. Glasses. A sunburned neck. Mischievous eyes. Pens in his pocket. We had talked twice on the phone before that, too—I was interviewing him for an article I was writing—and it was during one of these conversations that he mentioned he would be driving to the International Conference on the Monarch Butterfly in Morelia, Mexico, the following month, looking for butterflies along the way and doing some research, and invited me to come along.

This would be my second trip to Mexico to see monarchs. The first had been three years before, when my daughter was nine months old. That was how I would always remember the trip, with a certain amount of distance, as if I had been watching myself there: a woman in a foreign country with a small baby in her arms. We had been at a meeting, my husband and I, and at the end of it, as a kind of reward, we were to be taken into the mountains to a monarch butterfly preserve. Those words, *butterfly preserve,* meant nothing to me. I could not make them into a coherent image the way I could, say, Walt Disney World, where I had never been, either, or Glacier National Park, or Victoria Falls. What would a butterfly *preserve* look like?

We took a bus, and then a truck, and then we walked. At ninety-five hundred feet, where the climb began, the air was not so thin that you noticed, yet, how high you were. Other things were more obvious and would have taken your breath away even at sea level: the skinny little boys, for instance, who were selling things—recapped bottles of beer and snapshots of clustered monarchs and handkerchiefs embroidered by their mothers or grandmothers or sisters. The handkerchiefs cost a quarter, and though they were made by hand, all of them looked alike: a white cotton square with scalloped edges and an orange-and-black monarch butterfly sewn into one of the corners. That was the other thing that brought me up short: the butterflies. They were underfoot. I was used to seeing butterflies in the air, or on flowers, but there, at the entrance to El Rosario, thousands of wings torn from their bodies lay in the dirt. They were like cairns in the forest, pointing upward, and so we climbed, my husband, our daughter, and I, and the little boys fell away, and I could hear myself breathing, and my heart in my ears, and when I looked up again, what I was seeing made so little sense that I turned it into something else, something I understood—autumn leaves, falling through the air. That was what it sounded like, too. Millions of leaves, ripped and ripping from their moorings. The sound was overwhelming. It woke the baby in my arms, who opened her eyes to this sight. The three of us stood there, looking and looking, and gradually it occurred to me, gradually it registered, that though there were millions of them, they were not leaves at all, they were butterflies, monarch butterflies, the butterflies of my backyard. They were in the air, and so heavy on the branches of the pine trees that the branches bent toward the ground, supplicants to gravity and mass and sheer enthusiasm.

We moved on. As we hiked we saw even more butterflies, more than would seem possible, twenty or thirty million. Every available place to roost was taken. Even the baby became a perch. There were butterflies on her shoulder and shoes, butterflies in her hair. Somehow she knew not to touch them, and not to be afraid. We found a rock at the edge of the forest, and the baby and I sat down. The clamor of butterfly wings was as constant and irregular as surf cresting over rocks. I watched my daughter watching the butterfly resting on her shoelace, watched her reach down and wait until the butterfly crawled up the ladder of one of her fingers, climbed over the hump of knuckles, and rested on the back of her hand. She was completely silent, as if she had lost her voice. Her eyes were wide open, and so was her mouth, and for twenty minutes, maybe longer, the two of us just sat, eleven thousand feet up the side of a mountain, and paid attention. If I were a more religious person I would have called that place, and that moment, holy, or blessed. But my vocabulary did not typically include those words. Still, unbidden, they were the ones that came to mind.

AS A CHILD I collected rocks. Limestone, sandstone, mica, quartz—they all went in my box. Inside the box was a book with pictures of rocks, a field guide against which I would check my specimens, but what interested me most was how they felt in my hand, and their colors and consistency, not what they were called. The boy across the street collected butterflies, which he would pin to a piece of foamcore. Although I didn't have this word for it then—I was eight or nine—I thought it was morbid, which is to say that when my grandfather died the next year and I looked at his body laid out in the casket, rigid and perfect, it reminded me of those butterflies pinned to the board. The rocks, rattling around in my box, seemed more animate and full of possibility. Someday they would be dirt.

All these years later, I hardly remembered the difference between an igneous and a metamorphic rock. What I did remember was the single-mindedness with which I had picked through the woods behind my house, and the pure joy of finding something valuable enough to hold on to. It seemed reasonable to call this passion, and to think of myself—and everyone else—as a collection of passions. What this suggests is that it is not simply our ability to think, to be rational, that distinguishes humans from other species, but our ability to be *irrational*—to put stones in our pockets because we think they are beautiful.

All of us have experiences that could change our lives if we let them: love, offered suddenly, turning from the mantelpiece, as Delmore Schwartz put it. And that, oddly, was the way it was with me and the butterflies. Not love, exactly, offered suddenly, but a similar quickening of heart and desire—in this case a desire to *know,* if knowledge was not only information and understanding but experience. I could feel those butterflies tugging on my imagination as if it were a loose sleeve.

"YOU MEAN YOU agreed to spend a week in a car in a foreign country with a man you've never met?" my mother asked in disbelief the night before I left for Austin. She reminded me that I had a small child at home, and a husband. "Promise me you won't drive through Mexico at night," she said finally, and I did. I promised her up and down. And now it was near midnight and I was breaking that promise, doing something that only somebody who scored two sigmas more than a seventeen-year-old might find unremarkable.

"Are you worried?" Bill asked after a while, picking up, perhaps, on my body language, which spelled "tense" in marquee letters.

"Yes," I admitted. "Should I be?"

"I don't know," he said, but with conviction.

• • •

MONARCH BUTTERFLIES never fly at night. They can't. Once the ambient temperature drops below fifty-five degrees, they become sluggish, unable to flap their wings. The wings, which are commonly—and erroneously—described as solar panels, don't store energy but instead absorb it directly from the sun and air. Pick a monarch off a tree in the early morning and put it on your palm and it will sit there as if it were tame. Until the sun warms the air, the monarch is stuck paralytically, making it breakfast for certain steel-gutted birds. This is crucial because every autumn, monarchs do something no other butterflies do: they migrate unimaginably long distances. Monarchs born east of the Rockies typically go to Mexico. Those born to the west go for the most part to the California coast. They travel forty-four miles a day on average, but sometimes as many as two hundred, and all of it by day. Unlike songbirds, which often migrate in the dark to elude predators, monarchs are limited to flying out in the open when it is sunny enough for them, and warm enough, and not too windy.

"I THINK THEY are up there," Calvert speculated the next day, while the windshield wipers slapped at the morning drizzle as if it were something annoying, like bugs. The sky was gray, and Bill was narrating what we would have been looking at if we weren't looking at a thick curtain of clouds: big mountains, part of the Sierra Madre Oriental, mountains that rise eight and ten and twelve thousand feet into the air. In the foreground there were prickly pear and century plants and scrub grass, beautiful in their own way but hardly majestic. Calvert was sure the monarchs were on the move, but up high, above the fog—an unprovable hypothesis.

"Keep your eyes open," he said. "If you ever see a butterfly flying under these conditions—overcast, with no wind—it'll wreak havoc with all the existing theories." I understood implicitly that he would like this. Rules, even scientific rules, were anathema to him. But it wasn't going to happen; it was a biological impossibility. Thermoregulation was one of the few sure things that scientists knew about monarch migration. The rest—how the butterflies knew when it was time to leave their summer breeding grounds for their overwintering sites thousands of miles away, and how, navigationally speaking, they got to those sites—had stymied them for decades. And so had this: how did monarch butterflies from the eastern United States and Canada, *millions of them,* end up every year in the same unlikely spot, a remote and largely inhospitable fifty acres of oyamel pine forest ten thousand feet up the southwestern flank of Mexico's Transverse Neovolcanic Mountains?

This last question—how do monarchs find their way back to the same oyamel trees year after year?—remains one of the great unsolved mysteries of animal biology. Monarchs are not guided by memory, since no single butterfly ever makes the round trip. Three or four generations separate those that spend one winter in Mexico from those that go there the next. A monarch butterfly born in August where I live, in the Adirondack Mountains of New York State, for instance, will probably fly all the way to Mexico, spend the winter there, and leave in March. Then it will fly north, laying eggs (if it is female) on milkweed along the Gulf Coast in Texas and Florida before dying. The butterflies born of those eggs will continue northward, breeding and laying more eggs along the way. So will their offspring. By August another monarch, four generations or so removed from the monarch that left my land for Mexico the previous summer, will emerge from its chrysalis hidden among the raspberry canes and do the same thing. It will head south, aiming for a place it's never been, an acre or two of forest on the steep slopes of the Neo-volcanics.

All over the world—in the United States and Canada and Mexico, in Australia, in the United Kingdom, in Germany—people were puzzling this out. They were studying navigation, orientation, cellular structure, biogeology, ecology. All were doing science, and like much else in science, what they were doing was haphazard and unsystematic. Those of us who do not do science are often in awe of it, for it seems to possess a certain inherent power, the sort of power that comes from inalienable truths. For those of us who do not do science, science often seems to be the last bastion of unfuzzy logic, a place where the answers are clear-cut, a moral universe where there is a right and there is a wrong. But we fool ourselves—it's not like that at all. Science is ruled by human passions and limitations and creativity. Science is the story we tell ourselves, or are told, to make sense of the world of atoms and cells, illness and beauty, ozone and oxygen, the world in which we—collections of atoms and cells—find ourselves.

Every fall monarchs pass through my yard, and though I know where they are going, no one can tell me for sure how they get there. Maybe it doesn't matter; so much of nature happens in the background, without most of us paying it any attention. But paying attention is part of our animal heritage, a link to deer and dolphin, owl and bear. And that, perhaps, is where the itch of curiosity begins, and the ongoing attempt to scratch it.

People have been paying attention to the monarch for hundreds of years, and most of them, unlike Bill Calvert, have not been professional scientists. They have been drawn to these insects because their behavior has seemed so unlikely, so amazing, so nearly heroic. The monarchs' presumed heroism, in fact, has been the subtext for much of the fascination. In the process, data have been collected and then organized into a narrative. Ultimately, that narrative didn't have to be true, but like all stories, it had to *seem* to be true.

Bill Calvert was one of the storytellers, as well as part of the story itself. As an undergraduate at the University of Texas he studied philosophy, and as a graduate student there, zoology. His dissertation was on butterfly feet, how female butterflies find their host plants. He started looking at monarchs on a lark because he was stuck in Massachusetts one winter and wanted an excuse to go south. He went to a talk given by Lincoln Brower, then a professor at Amherst College and the world's leading monarch scientist. Brower needed butterflies for his research, and he put Calvert up to flying down to Texas to collect some. It was the beginning of a loose collaboration that continues to this day.

In monarch circles, which are bigger than one might suppose, Bill Calvert is something of a legend. It's not just his reputation as a cowboy entomologist, a guy who sleeps in his truck in pursuit of monarch butterflies and has more field notes and more data than he'll ever be able to write up— though these are part of it. What makes him a legend is that almost twenty-five years ago Bill Calvert figured out, based on a couple of clues in a *National Geographic* article whose authors were trying to keep it secret, where monarchs from the eastern United States and Canada spend the winter.

"I had a friend who was a librarian," Calvert said, "and she gave me a bunch of maps. There were two clues in the *National Geographic* article, that the butterfly colonies were at ten thousand feet and that they were in the state of Michoacán. When you put those two features on a map, there were not very many choices."

Calvert and three friends borrowed a truck and drove to Angangueo, a mountain town that was home to a silver-mining company once owned by the Guggenheim family. He was carrying a picture of a monarch butterfly, and when he showed it to the mayor of the town, the mayor became very excited and began to talk about a butterfly roost high in the mountains, a place called Chincua. It was the last day of 1976. Bill Calvert called Lincoln Brower and told him this. The next day Calvert and his companions found the butterflies on a ridge above Zapatero Canyon.

Bill Calvert gave up his postdoc on tent caterpillars. He bought lots of maps and started looking for place-names with the word *paloma*—"butterfly"—in them. He mounted expedition after expedition, discovering seven more colonies, roaming around Mexico on National Science Foundation money. Nearly twenty-five years later he was still roaming. He had a wife and a son back near Austin, but the marriage was breaking up. He was too restless—or she wasn't restless enough. "You pretty much have to be retired to do this research, wandering around, looking at things," Bill Calvert said. "You can spend all your time traveling around and not getting conclusive answers. So that leaves people like me to do it."

IT WAS NOVEMBER 6. According to the biological clock that unwittingly winds us all, the butterflies would be approaching their winter home. The door would be ajar, and they would be streaming over the threshold, marathoners tired from their long journey, eager for a patch of bark, or the branch of a tree on which to rest. Where Bill and I were, not far past the enormous concrete globe with a black line girdling the twenty-third parallel that marks the beginning of the Tropic of Cancer, there were pecan groves, row after row of them, and a carpet of yellow flowers. The rain had stopped, and though the day was bright, it was hazy. We hadn't seen a single monarch all day.

"If we're going to see monarchs today, this is the place," Bill said, stopping on the side of the road by an irrigation ditch. This is the kind of place they love." He pointed to the trees, which were bowed over the stream. "They just love these." We got out of the truck and began looking, searching the sky and the tree limbs and the water itself. We stayed maybe five minutes, as hopeful and enthusiastic as if we had never seen a monarch butterfly before, as if it wasn't the most common and best-known butterfly in North America. There is something self-preserving about the natural world—its ultimate adaptation—so that what is familiar and expected often seems new, over and over again: snow in winter, robins in spring, leaves turning in the fall. Bill Calvert has probably seen more monarchs than any other person on earth, twelve or fifteen or twenty million in a single frame, yet here he was, excited to maybe see one, right now, in this place. But he didn't. There weren't any. And he was disappointed.

"Let's go catch up with them," he said, so we got back in the truck and continued along Route 101. There were tapes on the dashboard—Gordon Lightfoot, Vivaldi, Bach fugues—but we drove in silence, looking, looking, looking, until it hurt to look so purposefully, at least for me.

"Over there," he said fifteen miles later, when we had begun the climb out of the lowlands into the mountains, and there it was, a single monarch, flapping its wings athletically, flying across the road. And, "There," again. Three more monarchs coming off the ridge to our right, heading southeast. And then more coming right over the truck, crossing the highway, sinking down toward the valley below, disappearing. Calvert stopped the truck and gathered up his binoculars, his tape recorder, his compass, and his global positioning unit.

"One at six feet at two-oh-five degrees," he said into the tape recorder after holding up the compass to take the vanishing bearings of the butterflies as they dropped to the valley. "Powered flight," he recorded, meaning that they were not gliding but were flapping their wings. "One traveling at ten feet at two-oh-five," Calvert called out. The number 205 referred to the monarch's azimuth, its direction with respect to magnetic north. In this case the monarchs appeared to be flying south-southwest. Calvert unsheathed the global positioning device and placed it flat on the ground, aiming it upward to beam a signal to a satellite passing overhead. "Let's find out where in the world we are," Bill said, turning it on.

The answer didn't come immediately. It reminded me of one of those Magic Eight Ball toys to which you direct a life-defining question ("Will I pass the math test?" "Will I find true happiness?") and wait expectantly while the answer floats into view ("It is too early to tell"; "Try again later").

The numbers started to drift in: 23 degrees, 23.25 N; 99 degrees 29.37 H. 4704 feet. To me they were less telling than what I could see: mountains as far as the eye could travel—big, imposing mountains that rippled like an inland sea, all the way to the horizon. We waited. Butterflies passed close by. Thirty monarchs in fifteen minutes.

"They're very patchy," Bill said when we were back in the truck. "I suspect ridges have something to do with it. I suspect that wind currents do, too. But it's hard for us to read the wind." Another two monarchs worked their way over us and dropped out of sight in the valley. Then three more. Then nothing. We had caught up with the monarchs, but we couldn't follow them. There was no road where they were going, so we moved on, scanning the sky, focusing on the foreground and the middle distance, but we saw none. We had lost them.

More than that sense of loss, I don't remember what I was thinking. We were coming to a town, Tula, where I hoped to find a phone to call home, so maybe it was that. But suddenly the brakes were on and we were making a dusty U-turn, and Bill Calvert was pointing to a stand of willows ten yards from the road, encircling a muddy pond. "Whoa," he said. "There's a roost. Probably two thousand monarchs in there." We picked our way over a barbed-wire fence for a better look, passing a great blue heron arrayed in its winter whites, and paused to admire a vermilion flycatcher, a bird so radiant I had to fight the urge to squint. Compared to this, the monarchs' orange and black markings, and especially their dried-leaf appearance when their wings were folded as they roosted, might have seemed drab. But the monarch itself did not. I had sometimes heard lepidopterists refer to the monarch's charisma, to its character, and as I stood in that swamp, looking up at the monarchs resting in the branches overhead, I knew exactly what they meant. These less-than-a-gram creatures had flown, most of them, nearly two thousand miles. They had almost made it. They seemed . . . admirable. Bill Calvert got out his equipment: a ruler, glassine envelopes, a digital balance, a tattered net, extension poles, duct tape. His subjects had arrived: he was going to do science.

WHEN BIRDS MIGRATE, they do so primarily because of food. Winter comes, and mosquitoes and berries and other food sources dwindle or become less accessible. Birds fly south, and the landscape becomes one big commissary. This is oversimplified, of course, but even schematically, what birds do is nothing like what butterflies do. Monarchs do not leave their northern breeding grounds because the flowers have withered. They leave for the same reason the flowers wither: the climate changes. The monarch butterfly, which is, genetically, a tropical species, cannot survive subfreezing temperatures. And when monarchs are wet, they are even more vulnerable. If they are going to reproduce, they have to move to a more hospitable place—or, as is really the case, a less *in*hospitable place. At ten thousand feet, the Neovolcanics are not the Bahamas for butterflies; the overwintering sites are not warm. Rather, they have the right microclimate for monarch survival, warm enough so the monarchs don't freeze and cool enough so they don't drain their finite supply of energy, the lipids stored in their bellies. Monarchs spend an average of 135 days at the overwintering colonies, days of entropy when food may be sought but is not much available.

But they need food; they need energy, both to fly long distances and to survive the winter. Intuitively one might expect the butterflies to bulk up in the north, the way we might fill up the gas tank before driving cross-country. The problem is that a loaded gullet may actually require more energy to transport. And it may cause drag. So the question of when a monarch obtains its winter food supply is an important one, both because it may suggest how the butterflies find their way to Mexico (do they, for instance, follow the asters and the black-eyed Susans?) and because it may have implications for conservation (what happens if wildflowers are replaced by roads or subdivisions or wheat fields?). Besides, it is just plain interesting: science for science' sake.

This last, more than the others, appealed to Bill Calvert: the questions, one begetting another—no; begetting many others. We would drive, and I would ask Calvert, who has devoted his life to studying monarch butterflies, how high monarchs flew, and if they followed corridors of wildflowers when they headed south, and if predation was greater during migration or remigration, and invariably he would smile and tell me that I had asked a good question and say, "But that's the thing, no one knows the answer." In my knapsack I was carrying around a book called *The End of Science,* about how scientists were closing in on a unifying theory to explain *everything,* and it seemed pretty clear to me, in talking with Bill Calvert, that the physicists were going to be able to tell us how the world worked, and we still wouldn't know how a single monarch butterfly found its way from Canada to Mexico, or the answers to the hundreds of questions raised by its flight.

THE BUTTERFLIES at this particular roost were fifteen, twenty feet up—too high to grab with a regular butterfly net. Calvert rooted around and found a long stick, which he taped to his extension pole. Then he taped this gangly arm to the handle of his net. It almost reached—he was going to have to jump. The monarchs, meanwhile, were sitting ducks. A few were milling about in the air, but most were lined up, thorax to proboscis, along the branches. Calvert made one practiced swipe and nabbed about thirty butterflies, who seemed astonished to find themselves suddenly crowded into a green mesh funnel and who tried frantically to escape. But it was impossible. Calvert had deftly tossed the bag over itself, effectively sealing it. Nothing could fly out.

We got to work. The scale was set up and calibrated, and I started a log sheet, with columns for time, weight, sex, and condition. Bill sat cross-legged on the ground, pulling butterflies from the net one by one, measuring their wingspan, and noting how tattered or not they were and which sex. While I wrote all this down in the log, Calvert would fold, then stuff, the monarch into a two-inch glassine envelope and lay it on the scale. "The envelope is so it won't hurt itself," he explained. It prevented the monarch from flapping wildly. Once it was in the envelope, he could lay it down on the scale and get its weight. I recorded that, too.

"These butterflies are very heavy," Bill observed. "You'd think they would have a harder time with a heavier load."

"Maybe the load supplies extra energy," I suggested. It seemed as good a theory as any.

Bill Calvert considered this for a second. "There are many trade-offs in a butterfly's life," he said, smiling.

Calvert had once thought that monarchs started their journey south as lightweights and didn't begin to add weight until they reached Texas. They would lose some crossing the desert and then begin to nectar heavily once they were in Mexico, in anticipation of the winter to come.

"But now I'm not so sure," Bill said. "I contracted with a woman in Milwaukee to capture monarchs and send them to me overnight, and I weighed them, and they weighed a lot less than the butterflies I was capturing in Texas. But then I had her weigh them before she sent them from Milwaukee, and I weighed them the next day—they were sent FedEx— and they were considerably lighter. I think the definitive paper on the use of nectar during migration remains to be written."

TWO BURROS WERE tethered to the fence opposite us, and there were two women there, too, waiting for a bus. But the bus was nowhere in sight, and before long the women asked us what we were doing, and we said the words "*Bio-logico*" and "*Mariposa monarca,*" and they grunted and shook their heads and smiled at us and kept watching. Watching and smiling and pointing and laughing. We *were* pretty enter-taining, especially when a butterfly would push its way out of an envelope and we'd flop after it, trying to get it back.

As the women watched us, I watched Bill Calvert sitting Indian-style in the dirt, measuring and weighing the butter-flies and assessing their condition on a scale of one to five. Most of the monarchs were in good shape, all things consid-ered, fresh and untattered. Focused as Bill was, I noticed that his affect was that of a child—happy, engaged, fully present. He reminded me of my daughter at the edge of our pond, sorting rocks, digging for turtles, catching newts. Children have enthusiasms and adults have passions, I thought, and though it sounded good, Bill Calvert seemed to contradict it. The answers mattered to him—that was a given—but getting at them mattered even more.

"The need of the age gives its shape to scientific progress as a whole," I read months later, in Jacob Bronowski's classic meditation *Science and Human Values.* "But it is not the need of the age which gives the individual scientist his sense of pleasure and of adventure. . . . He is personally involved in his work, as the poet is in his and artist is in the painting. Paints and painting too must have been made for useful ends; and language was developed from whatever beginnings, for practical communication. Yet you cannot have a man handle paints or language or the symbolic concepts of physics, you cannot even have him stain a microscope slide, without instantly waking in him a pleasure in the very language, a sense of exploring his own activity." It was an old book. I had underlined the passage twenty years before. But it described precisely what I was seeing while I was watching Bill Calvert.

THE BUS CAME, and the women waved and turned and got on it. It shambled on, dust rising behind it. The heron paced the bank; our log list grew longer. The butterflies were healthy and fat. Their trip appeared not to have taken a toll. A young man rode by on a bicycle, doubled back, looked at us, rode by, and doubled back again—like a yoyo. Cheered by the bonhomie of our encounter with the two women, I decided to clue him in on what we were doing on his next pass by. "*Mariposo,*" I shouted loudly as he rolled past. "*Mariposo!*" Bill Calvert looked up at me and smiled. I smiled back, happy to have explained our mission with such ease and precision.

"You'd better hope he doesn't come back this way again," Bill drawled as he casually stuffed a monarch into an envelope. I looked at him curiously. "You just called him a faggot," he said.

THAT NIGHT WE WERE stopped by the police. Or maybe it was the army; it wasn't obvious which. There were sixteen of them, in knee-high black paratrooper boots and black pants and black sweatshirts, and they had AK-47s and Uzis. It was dark. We were in the desert and it was late. We had gotten cavalier about where we went, and when. We had been taking our time all afternoon, inching our way along the side of the road just past Tula, looking for roosts. Every couple of trees Bill would say, "Over there" or "A thousand on that one," and "That will grow by a factor of ten by nightfall." He said this so confidently that I wrote it down in my notebook as if it were fact, not prediction.

The land outside Tula was arid; rain seemed a memory. The ground was cracked and it curled like smoke when the wind blew. The wind was blowing. Monarchs were dropping out of the sky. Those that were flying at tree height were being tossed around like falling leaves. They were fighting back, treading the air by pumping their wings, but often blowing backward. "Golly," Bill said to me, "there are a lot of them." And then, to his tape recorder: "The last five-mile segment there were at least one hundred butterflies."

We drove due south. When the wind let up, the monarchs escorted us, flying straight along the edge of the road as if they were pedestrians on a sidewalk. But then, as if there were a sign, or a crossing guard, or a traffic light, they all turned at the same spot and went to the other side of the road. We stopped the truck and got out and looked up. The trees were teeming with monarchs. I followed Bill across the road and saw him enter a grove of huisache trees and drop to his knees. There was little understory here, but enough to get my legs full of cactus thorns. Bill was rooting around the leaves at the base of one of the trees. "Mouse cache," he said as I walked up behind him. He pointed to the pile of leaves, only they weren't leaves, they were monarch wings—hind wings, forewings, left wings, right wings. Wings, no bodies. "The mice eat the bodies and leave the wings," Bill said. I poked around with the tip of my boot. There were hundreds of them.

When we crossed the road again to walk back to the truck, the sun was going down. Not one to pass up an opportunity, Bill got out his net and his scale and his ruler, and I started a new page in the logbook, sitting on the hood of the truck.

So we were late, and crossing the chaparral in darkness. Not late for anything in particular, though I guessed there were chicken mole and Dos Equis and a marginal hotel room not far ahead. The road had turned bumpy, and then there was a detour sign, and we followed it, though it took us off the pavement and through dried streambeds and gullies that the truck strained to climb. The truck, which was already low to the ground, bounced on its shocks like a pogo stick. Bill gripped the wheel and fought to keep us upright. All of the things in the truck bed, Calvert's carapace, really—the sleeping bags, gallon jugs of water, the *Random House Dictionary,* a Spanish dictionary, woven mats, nets, our gear, his boots— crashed into one another and into the windows. They were timpani to the engine's tuneless melody. We were gaining altitude, little by little. Outside of the narrow band of the high beams, everything was black. It was as if the night were a well and we were submerged in its ink.

They must have seen us, then, long before we crested the last hill. They must have seen us dipping into each ravine and heard us pulling out. Their lights were riveting, like klieg lights when you're standing on a stage, and there was no choice but to stop. They made a ring around the truck, each one pointing a gun. One of them opened the door and motioned to Bill, with a wave of the barrel, to step out. I was like a monarch in the morning, unable to move. A gun, though, can motivate you. The handle of the door on my side turned, and when I looked to see what it was, I saw the midsection of a man with a gun trained on me.

We showed them our papers—first our passports, then the ones that said we were going to a conference on monarch butterflies. I was careful not to say the word *mariposo*. I was careful not to say anything. The men with the guns handed back the papers and opened the hood and peered inside with flashlights. They took off the hubcaps and looked in there. They dumped out our trash and rifled through our books. Bill and I didn't talk. I knew what he was thinking: if they found the glassine envelopes and the digital scale, we were in big trouble. It occurred to me that we had picked the perfect cover for running drugs.

The men with guns thought so, too. They checked out our field glasses, the tape recorder, our cameras. They looked behind the heating vents. They pulled up the floor mats. The lights of a city, maybe two miles away, winked as if they were in on a joke. One of the policemen was wearing a U.S. Army surplus jacket that had once belonged to a soldier named Olson. They started in on our duffel bags, feeling them up and down as if frisking bodies. The scale was back there, too, in a knapsack stored inside a backpack. The backpack had many pockets. It was freezing outside.

"*Basta,*" said one of the men. How long had it been? Forty minutes? Fifteen? The others lowered their guns. The one with the army surplus jacket nodded at us.

"*Mariposa monarca,*" he said.

Back in the truck, driving again, the heat was on, but I couldn't stop shivering.

"Bet you didn't know studying butterflies was such a dangerous occupation," Bill said.

Chapter 2

OF COURSE we kept driving at night. There was no time during the day. We'd go three miles and stop to scan the sky or poke around for roost sites or evidence of roost sites (disembodied wings, half-eaten thoraxes), go another few miles, hop out of the cab, do it again.

We were in dry, poor country. The houses were made of concrete and tin, or sticks and mud. Smoke rose thinly from makeshift chimneys, and threadbare clothes hung from wash lines. Water was carried in. The word that came to mind was *abject*. But the word *poverty*, which is typically twinned with *abject* in such circumstances, seemed far too modern. This place was preindustrial, sixteenth-century, with not a power line or a phone cable or a car in the yard to be seen. The yards, in any case, were scars of earth where nothing grew. Hunched old women with bundles of spindly logs slung on their backs walked the roads, bringing home cooking fuel. Rib-skinny dogs trotted alongside them, scavenging for anything remotely edible, while turkey vultures patroled the sky, scavenging the dogs. This was not the Mexico of the off-the-beaten-track tourist guides. There *was* no track.

Outside Ahuacatalán, in the dusty heat, we saw monarchs high in the sky and stopped to watch. I was conscious of our binoculars, and global positioning device, and tape recorder, and cameras—of how absolutely rich we were, in relative terms. As if reading my mind, a drunken young man in black jeans and a cowboy hat came reeling up the road, mumbling to himself, carrying an unsheathed machete that he twirled absently in his hand. He stopped nearby and stood at the edge of the road, peering into the same distance we were peering into, trying to see what we were seeing. The machete impressed me into silence, and I stood staring skyward, as if I could will myself up there, away from that blade. And maybe I did. After a few minutes the man wandered off, though I dared not lower my field glasses to see where he'd gone.

"I guess we'd better get going," Bill said finally, when he had seen enough, and then he proceeded to walk past the truck and continue down the road as if he'd forgotten where he'd parked his vehicle.

"It occurred to me that we might find a roost," he said when I caught up with him, fifty yards later.

But we didn't, not then—not until the next morning, when, driving out of Tequisquiapán, we took a wrong turn down Avenida Cinco de Mayo, which dead-ended at a stream decked with cypress trees. "This is perfect for monarchs," Bill said, pointing to the water and the trees. He moved his finger two degrees to the left. "And there they are."

And there they were—a pair of monarchs chasing each other five feet above the middle of the streambed. We trailed them like spies, hanging back a few feet, trying to stay out of sight as we picked our way along the water's edge and were led unseen to one roost site, and then another, till we counted four of them in all, each with about a hundred butterflies.

WHEN PEOPLE FOLLOW the laws of a nation—when they pay their taxes and stop at red lights and respect others' privacy—the infrastructure that lets us live together is transparent, and no one really notices it. The laws of nature are different. When the natural world conforms to them, or at least when it conforms to certain patterns, one glimpses, and understands—*if understanding is a feeling*—the origin of magic.

That morning was magic, even when, an hour later, we had progressed no more than two miles and stood on the side of a busy road, and not a single monarch of the dozens we were seeing was going in what was supposed to be the "right" direction—that is, the direction that would lead it to its winter habitat. (The wrong direction, meanwhile, would ultimately send it back to the United States.) Calvert was unperturbed.

"It seems to me that I've run into this before," he said, more to himself than to me. "Their initial behavior in the early morning is east, toward the sun, and then they warm up and head southwest."

Then we were heading southwest ourselves. At most, the butterflies were intermittent. We stopped near San Juan del

Rio at Comercial Mexicana—a Mexican version of Kmart—
to stock up on bottled water, but we never made it into the
store. Calvert sensed that there were monarchs overhead,
sensed them the way a dowser smells water, and though I
couldn't see them myself, I wasn't surprised when I looked
through a pair of binoculars and saw them skipping across the
very top of the optical range. They were like stars in a cloudy
night sky, only vaguer. "Let's take some azimuths," he said,
so we did. The butterflies were going the "right" way, and
so were we. That was the day when, at long last, we entered
Michoacán, the state where most of the monarchs overwinter
and where the North American Monarch Butterfly Conference would begin in Morelia, the capital city, the next day. It
was an uneventful crossing through a scorched and scrubby
desert, but it meant we were in range of the monarchs' winter
home. Flat though it was where we were, we could see tall,
rugged mountains in the distance and began to gain altitude
ourselves.

"I think the butterflies may use those mountains as bea-
cons, to guide them in," Bill Calvert said casually, as though
that thought had just popped into his head. But I had heard it
before, in different versions, all of them his. We were taking
the high road ourselves, on the spine of the Sierra Madre
Oriental, then dropping into the valleys, because of this very
notion—the idea that monarch butterflies might use these
mountains as a focusing mechanism to set them on a narrow
path leading to the preserves.

"A butterfly born in Minnesota and one born in New
York State end up in the same intermontane valley because
of this focusing device," he said. "When they start out they're
spread two thousand miles across the continent, but when
they get into Mexico they're condensed into just fifty miles.

Wherever they join the mountain ranges of the Sierra Madre Oriental, they turn and follow it."

That was the theory—the only one, Calvert said, that accounted for longitude in a monarch's migration. The problem was that the hypothesis was basically unprovable. Calvert ticked off a list of questions that could be answered, it seemed, only by radio telemetry: "Where do monarchs hit the Neovolcanics? How accurate are these creatures, anyhow? How do they know when to stop? Are they really directional? There's some evidence that suggests that they're not." But so far, sticking a radio transmitter on a monarch butterfly had not been an option—the devices didn't come in that microdot size, and even if they did, using them would probably feel like cheating to an inveterate field biologist like Bill Calvert, a guy who made his own furniture, poured his own concrete, raised his own bees.

We pushed on, and the mountains got bigger, and as they did, Bill began looking at them and not at the sky, and it became clear that they were totems for him—totems to an earlier, less complicated life.

"Boy, I've spent so much time wandering these roads and climbing those mountains, looking for monarchs," he'd say. Or, "People would ask me, 'What good are monarchs?' I hate this question. Basically it's a matter of aesthetics. Either you love these creatures and this phenomenon, or you don't."

For more than two decades they had been the one constant affection in Bill Calvert's life: a child's life wedded to adult ambitions. Or maybe it was the other way around, an adult's life wedded to a child's ambitions. Either way it was seductive, this search of his, this responsibility to nothing but the questions. Where I came from, it was the answers that mattered most: had my daughter's cough cleared up, had the doctor called in the prescription, had this winter's firewood been cut and stacked, had the paycheck arrived, was the phone bill paid? As I drove along with Bill Calvert, I sometimes made mental lists of these for the rare times when we'd find a phone and an operator willing to place a call across the border. For the most part, though, my questions became simpler and less answerable: "Where are the monarchs?" "Will we see monarchs?" These questions could be enough, I was learning from Bill, to build a life around.

TEN MILES FROM Coroneo we pulled over to get another set of azimuths. The monarchs should have been closing in on the overwintering sites, which were to our west, and the azimuths should have reflected this. *Should* have. But didn't. The butterflies were going south. The monarchs were flying at about a thousand feet—high, barely visible. Calvert wondered if they were heading for San Andreas, a wintering area that in recent years had been ravaged by fires and logging. Who knew? We were in farmland that ended in a wall of mountains. Gunshots pop-popped somewhere close by. Unfazed, Bill kept talking into his tape recorder, giving the azimuths. "It's just a rifle," he said, turning away from the machine to reassure me—saying, in effect, that we were hearing merely the call of an unremarkable bird. And as it turned out, he was right. In a nearby field, two boys were shooting at crows.

SO WHAT IS DANGER? For me it is a feeling, sensual and percussive and paralytic. For a butterfly it may be this, too—

we can't begin to know—but it is also a constant condition, like weather. Indeed, weather itself presents one of the greatest dangers in a butterfly's life, and particularly that of a migrating monarch butterfly, which covers thousands of miles through uplands and lowlands, across water and along coasts, in its journey south and then north again. Too cold and the monarch can't fly, might freeze. Too hot and it gets overheated, can't fly. Too hot and there might not be enough water. Too much wind, grounded. Wind from the southeast, stalled. Wind from the west, blown seaward. Hurricanes. Tornadoes. Snow. All present dangers, and not just to monarchs, but to their habitat as well. Of the 106 known species of milkweed, only about a dozen are used by monarchs as sites on which to lay their eggs. The milkweed is essential, for it provides the cardiac glycosides—the poisons—that are ingested by monarch caterpillars and that in turn make monarch butterflies poisonous to most birds. Although perhaps hardier than the monarchs themselves, milkweed plants nonetheless need adequate rain, but not too much rain, and adequate sunshine, but not too much sunshine, in order to grow. They are weather-dependent, too. The absence of suitable habitat for breeding, migrating, or overwintering breaks a link in the chain and puts monarchs at risk.

So do predators, which are numerous: wasps, fire ants, earwigs, aphids, rodents, birds, and, in their own way, people. Cows like to eat monarchs: Mexican farmers used to bring cattle up to the overwintering grounds and smoke the butterflies to the ground, where the cows would eat them by the thousands. Mice like monarchs. Once I was raising a newly hatched monarch whose crumpled right forewing made it unable to fly. Day after day it would sit in my kitchen sucking up sugar water from a saturated sponge. It was there as usual on a Thursday night and then, suddenly, not there the next morning—totally absent unless you counted the mouse droppings on the counter not far from the sponge. And raptors, which you might assume would have bigger fish to fry, like—as in "find tasty"—monarchs, too. Watch a thermaling sharp-shinned hawk through binoculars and you're likely to catch sight of a thermaling monarch in the same funnel of wind, but only for a minute if the hawk happens to be hungry.

To call people predators is perhaps a stretch, but only if you assume that predation requires intent. For the most part, people are monarch butterfly predators not by design but by default, as when they mow a highway median strip at the wrong time and eliminate thousands of acres of accessible milkweeds; or when they plant genetically modified corn infused with a toxin aimed at killing corn borers, which also, through its pollen, kills monarch caterpillars; or when they spray crops with herbicides and pesticides; or when they cut down trees in the Mexican overwintering sites, thinning the protective canopy and altering the microclimate it creates, which together allow the butterflies to survive both the cold and the breakfast-time raids of orioles and other birds of this particular emetic appetite. It's the same old ecological story: everything is connected.

The fact that the process is circular, not linear, poses its own danger, too. What I mean is this: it's easier to identify problems that arise in causal relationships and then to address, if not remove, them. If your boyfriend hits you, for example, you can leave him and no longer be in the path of his blows; it may not be that simple a relationship to leave, but you understand what you have to do. With monarchs, however, there are many potential "batterers," few of whom actually mean to hurt the butterflies. The loggers in Mexico may be thinking only of the money that a truckload of oyamel fir trees will get them, the corn it will buy or the heat it will furnish; monarchs may never enter into the calculation. But the loss of the trees puts them in jeopardy. The farmers in the midwestern United States who plant genetically altered corn may be thinking only about increased crop yields, not how far the pollen travels or whether monarch caterpillars will ingest it and die. The road crews in New York State may be thinking only about driver safety when they raze the weeds and grasses along the highway, not realizing that in so doing they are eliminating a major food source and breeding ground for migrating monarchs.

And if culpability is difficult to assign, changing these practices may be even harder. How easy it is to feel insignificant when you're part of something so much bigger than yourself, to just go on about your business as if your particular actions had no consequences, or even as if your particular aggregate actions had no obvious consequences—if you're a farmer in Iowa, say, or a tree cutter in Michoacán, and you know that even if you personally don't do that thing that is destructive to monarchs or to monarch habitat or to monarch larvae, someone else, somewhere else, might. And then who will know whose fault it is that the web is coming undone?

MILES FROM the capital we lost track of the monarchs again. We were near Acambaro and saw none, and continued to see none as we at last gave up the back roads for the highway and began to approach the outskirts of Morelia, a thriving, well preserved colonial city. No butterflies—they were off our radar screen completely, and I was disappointed. I had read accounts of monarchs' streaming into the overwintering sites by the thousands, so many monarchs that the sky was a ribbon of orange and black, or the orange and black were a thick curtain obscuring the sky. I had been expecting to see drama on this trip—the *Mutual of Omaha's Wild Kingdom* kind of drama, not the *Midnight Express* drama of police checkpoints—but nature was not compliant. The city began to unfold, its sand-colored buildings and cobblestone streets, and the familiar hum of commerce, drawing us in. There were signs for cell-phone vendors and Pizza Hut and Ford trucks. We had returned to our own commercial moment, and it was one in which the questions "Where are the monarchs?" and "Will we see monarchs?" were beside the point. "Where's the cash machine?" and "Is there valet parking?" were more like it.

But not for long. After checking into the Gran Hotel and making those inquiries, Bill and I walked over to the convention center to sign in. And there were our fellow conferees, congregating in the hall, one of them talking about how, when he was driving here from Mexico City earlier in the day, there had been so many monarchs crossing the highway that traffic had come to a standstill.

"We were in the wrong place at the wrong time," I suggested to Bill Calvert after hearing that.

"Maybe," he said in his enigmatic way. Which I took to mean, "Maybe not." It was a pretty compelling image, those butterflies halting traffic, he seemed to be saying, but the story that our trip would tell, once it had been added to the trips he had taken before and planned to take in the future, would have more depth, more uncertainty, more truth to it. Our *not* having seen monarchs was as revealing, and ultimately as dramatic, as our seeing them would have been. And a week on the road in a messy truck with a laconic man who had studied philosophy before turning to bugs had me believing him.

NOT THAT the people milling in that hallway were easy to dismiss; their names were among the pantheon of monarch biologists and ecologists and economists. Steve Malcolm, an Englishman now settled in at Western Michigan State University, and Myron Zalucki, an Australian, had both done pioneering work on host plants. David Barkin, an American economist who lived in Mexico, was there, as were Brooks Yeager, an assistant deputy secretary at the United States Department of Interior, and Steven Wendt, the head of the Migratory Bird Department of the Canadian Wildlife Service. Later they would be joined by Lincoln Brower, perhaps the preeminent monarch biologist in the world, as well as Karen Oberhauser of the University of Minnesota and Orley "Chip" Taylor of the University of Kansas. To anyone with even a passing familiarity with D-Plex, the e-mail discussion group sponsored by Monarch Watch, a monarch butterfly tracking program run by Chip Taylor, all these names had resonance. So did those of Don Davis and Paul Cherubini, two amateur lepidopterists whose spirited and voluble contributions to D-Plex had gained them a certain notoriety, and who had made the trip to Morelia from Toronto and Sacramento, respectively.

Not everyone who was anyone in this world was there: Robert Michael Pyle, the author of almost every North American butterfly field guide ever written and an expert on western (United States) monarchs, was noticeably absent, as were the aged Professor Fred Urquhart of the University of Toronto, whose original efforts had led to the August 1976 *National Geographic* article that set Bill Calvert on his search for the overwintering sites, and Homero Aridjis, the celebrated Mexican poet who had grown up in Michoacán, written extensively and beautifully about the monarchs, and been greatly responsible for the 1986 presidential decree protecting certain overwintering sites from logging. But Bruce Babbitt, the United States secretary of the interior, would be flying in, as would his Mexican counterpart, Julia Carabias-Lillo, the head of the Secretariat of the Environment, Natural Resources and Fisheries. And there would be countless functionaries from both governments, and from Canada's as well, and representatives of NAFTA, the trade organization that had, perhaps unwittingly, chosen the monarch butterfly as its symbol. In many ways it was a more apt symbol than anyone might have first imagined: an insect that made its way from Canada to the United States to Mexico and back again demanded, in real terms, the kind of cooperative nurturing whose benefits would really accrue only beyond the realm of commerce.

"FACE IT," one of the biologists said to me later that night, when a bunch of us went out for dinner. "The idea that a monarch butterfly is going to get these countries to act in a less than self-interested way is a completely naive point of view. This whole thing is about money. A lot of it is going to get thrown around here." I took his point and kept my eyes open, and what was interesting, I soon saw, was how the basic truth of ecology—that everything is connected—was being leveraged by government types and representatives of NGOs and even some of the scientists to promote economic change in Mexico. So if by day two it seemed strange to ride the elevators of Morelia's Gran Hotel and hear endless conversations about isotope fingerprinting and nectar densities, it was even more disconcerting to observe a tiny insect's moving grown men and women to attempt, even on a modest scale, human social engineering. Or at least to talk about it.

"This is an area of great biological wealth and dire poverty," Julia Carabias told the hundreds of conference participants, speaking of the impoverished places, not far from where they sat, where the butterflies spend the winter. In the audience was a busload of people from those villages, campesinos in western wear with weathered faces who had been shipped in either (depending on how cynical you were) to state their own case against conservation efforts that would not also aid them or to justify those very efforts. "There is intense use of the soil. This turns into a very difficult problem. There are many conflicts of interest. The focus can't just be about the protection of the forest. We must preserve but develop the area so that living conditions are adequate."

Carabias, a very beautiful and charismatic woman, sat down to terrific applause. Everyone knew that what she was saying was rhetoric, but there was something about her presence that was reassuring. She appeared to be sincere, sympathetic, uncorrupt. For the most part, during her address the campesinos were sitting straight up in their chairs. She was talking about them, stating their case. I may have been projecting, but to me they looked hopeful.

And then Carabias went back to Mexico City, leaving the real work of the conference to her underlings, who had few of her charms, and the hopefulness began to dissipate like the bubbles in a half-capped bottle of soda. On the third day, when a small, gap-toothed campesino stood up during a question-and-answer session and said, "The support you have given us, putting us in a good hotel, frankly tires us," he was articulating a more general sentiment about substance and illusion. He wouldn't have minded staying in such place, in other words, if something useful was going to come out of it. (Bill Calvert, too, seemed just startled to find himself in such luxe accommodations, and anxious to get back to his truck.)

As Julia Carabias intimated, the Mexican situation was complicated. Because the bulk of the eastern United States' population of monarch butterflies and some of its western monarchs congregate in so small an area in Mexico, they are especially vulnerable to changes in that habitat. That was a given. The problem was not that the butterflies would face extinction if they lost this habitat; it was that their migration could be wiped out. Monarchs would survive in other places—on the West Coast of the United States, in Mexico and Florida, where there are resident, nonmigratory populations, in the tropics, in Hawaii, in Australia—but this phenomenal adaptation, this instance of a butterfly that behaved, in its seasonal long-distance commuting, like a bird, would be gone. Still, in a way it was a hard case to make convincingly, since it was not about species extinction per se. As Bill Calvert said, what it came down to, really, was a matter of aesthetics.

Meanwhile, there were people in those forested areas for whom aesthetics were as much a luxury as the little soaps and shampoos in the rooms at the Gran Hotel. "It is not fair that the forest farmers should pay the whole price for conservation," one of their representatives, Silvano Aureoles, told the group. "It is most important to know what people need."

"Sustainable forestry," the American secretary of the interior, Bruce Babbitt, said emphatically when questioned later on this point. "This is an absolutely classic example of the paradox of finding ways of living on the land for sustainability."

Still later, a Mexican peasant named Homero Gomez put it this way: "There is high illiteracy, high population growth, extreme poverty. The forest problems won't be diminished if these aren't addressed." His compatriot Dimas Salazaar agreed, saying, "The population is increasing, but the land and the water don't grow. The forest becomes smaller and smaller. We're going to eat each other." Here he grinned widely and laughed, and for a moment the whole thing seemed impossible and funny because it *was* impossible, so maybe we should talk about something else, like the grandkids or the weather. But suddenly Salazaar grew serious: "If nothing happens, we are going to take action. We are hopeful that something will come out of this, but if not, we'll take action. I am not saying this is the next Chiapas, but . . ." He stopped there. He knew what he was doing.

I had met Dimas Salazaar the day before, after he stood up in the middle of the gathering and said something unlike any other utterance that had been made in that cavernous room (which was arranged like a wedding, with a bride's side—the Mexican campesinos—and a groom's—the American and Canadian scientists and environmentalists): "To the scientists, if you can speak the language of the monarch butterfly, please thank them. Those little animals give us the opportunity to express ourselves as peasants."

In my notebook I had written down a brief description of the man who spoke those words, as much to help me track him down afterward as to remind me later what he looked like: "Stubby, middle-aged, tall white cowboy hat, tight jeans, bowed legs, dark leathered skin, silver teeth." But he wasn't hard to locate. Although he wasn't a big guy, he created around him a big aura. Over on his side of the room he seemed to be holding court, kidding around with other men in jeans and cowboy hats, all of which activity stopped when I approached and asked to talk with him. Dimas eyed me quizzically for a moment, then cleared a space at the table and invited me to sit down. "Me, talk to me up-front, I don't have to cover my face," he said, and he broke into a big smile and made a joke in Spanish that I missed but whose punch line seemed to have to do with me. Everyone relaxed, more campesinos joined us, and Dimas began to play the host, as if he were showing me around the land he farmed in Zitacuaro—not far, he said, "from the Safeway"—where he grew peaches and raspberries and corn and wheat.

Although Dimas owned a small parcel of this land—three hectares, or about eight acres—much of the land worked by him and his neighbors, many of whom were also his relatives, was communally owned. This was the legacy of Zapata's radical constitution of 1917, put into practice three years later, when Mexican land was divided up and distributed to the peasantry, and of the land reforms of President Lázaro Cárdenas a decade and a half later. The unit of administration for these communal lands was called the *ejido,* and its members were *ejidatarios.* Dimas Salazaar, in fact, was at the Morelia conference as a representative of the Alianza de Ejidos y Comunidades de la Reserva Mariposa Monarca. These were the communities whose land happened to coincide with that inhabited four or five months of the year by the butterflies. Dimas's own collective, the Ejido Francisco Serato, jointly held 313 hectares of land, one hundred of which were mountainside forest in the Chincua preserve, one of the biggest (in terms of the number of monarchs it harbored) and healthiest (in terms of forest vitality) of the overwintering sites.

AFTER DIMAS SALAZAAR had made his impassioned plea to the group at large, asking the scientists to thank the butterflies on behalf of the peasants, after he had thus gotten everyone's attention in a way that ran counter to the rampant (though not yet virulent) animosity of the day—typified by a complaint from one of the Mexicans that "since monarch science is in the hands of scientists, it is elitist"—he said something else, with equal conviction, that nobody missed: "We want to make a general request that the 1986 decree be reviewed, taking into account the owners of the forest." If the first part of his statement garnered great applause from the groom's side of the aisle, the second part received even more from the bride's. The decree to which he was referring was the one that Homero Aridjis had midwifed, and the reason so many *ejidatarios* were suspicious of Aridjis. The decree, they believed, was the cause of the worst of their problems: desperate poverty, chronic unemployment, hunger.

"Before 1986 we could work more freely in the forest," Dimas explained. "We would cut the trees reasonably and they would regenerate naturally. After the decree we had to do this clandestinely. The decree has meant that the police are involved. Then people pay off the police. Then it becomes a mafia. We don't hate the butterflies. What's happened to us has been done by men."

THE 1986 DECREE set aside five roosting sites, all in the Transverse Neovolcanic Belt, in the states of Mexico and Michoacán. Although the land was still nominally owned by the *ejidos,* the decree was specific: it designated both a core zone and a buffer zone in each of the preserves and described what the owners, *ejidatarios* such as Dimas Salazaar, were and were not allowed to do in them. The core zone was off limits for logging and farming; the buffer zone was not. In practice, what this meant for the *ejidatarios* was that a certain amount of their land had been effectively appropriated by the government and was no longer theirs to do with as they wanted. Logging was not completely banned, but it was, to a large extent, driven underground, creating a market for illegal trees. So for the *ejidatarios,* many of whom lived in unheated houses and subsisted outside the cash economy, taking the forest out of circulation compounded their hardships.

"In 1986, when the government created the five monarch butterfly reserves, we thought we'd get some benefits," Silverio Tapia, president of the Ejido Jesús Nazareno, told the conferees when it was his turn to speak. "But that never happened. Finally, in 1990, because our families were starving, some of us decided to log in the area without a permit. The government noticed and treated us like criminals. We went to jail because we took out some trees to feed our families."

· · ·

SO IT WAS the snail darter/spotted owl thing all over
again, a standard-issue environmental controversy that pitted
local residents and their livelihood against outsider conser-
vationists. Of course the threat of violence upped the ante,
and the dire poverty added extra color, as did the fact that
there were two languages (more if you counted the indigenous ones) and three countries involved, and at least one of
the countries had some money to throw around. Walkabout
money. Which is why the other refrain that could be heard
throughout the auditorium was that the money should go
directly to the *ejidos* and the *ejidatarios* and not to any middle-
men—not to the bureaucrats, not to the NGOs, and especially not to the Mexican government, where money had the
habit of disappearing before it got to where it was meant to
go. But no one really, truly ever believed it would happen.
No one, that is, with the possible exception of the man holding the strings to the biggest purse, Bruce Babbitt, and what
he had in mind was not exactly a direct deposit into the accounts of the *ejidatarios*.

"Buying up the land from the people who live there is a
very superficial way to deal with this problem," he said,
referring to a couple of proposals, some from the bride's side,
some from the groom's, to pay the peasants for their land, by
either buying it outright or renting it. "We've got to get into
the culture of these communities and find pathways toward
economic development. An ecotourism industry. That's the
kind of alternative development we should be investing in."

It was an attractive notion, ecotourism, especially the way

the secretary was envisioning it: small guesthouses run by local folk who would also promote and sell indigenous crafts; visitors coming all year round, even when the butterflies were not in residence. But there was no infrastructure in the region to support it, and the idea also failed to take into account the butterflies and their habitat, and the impact that increased numbers of visitors would have. Already, more than a hundred thousand people—most of them Mexicans—had passed through El Rosario, the first butterfly preserve to capitalize on people's desire to stand among millions of monarchs. But the dust that their feet had kicked up, as well as the concomitant erosion, trash, and water pollution they'd brought, had begun to worry some people, among them the Mexican biologist Benigno Salazar. "We need to determine the carrying capacity of the area," he told the Morelia conference a few days after Bruce Babbitt had gone home. Tourism, he made clear, was not a panacea; it, too, had to be regulated and limited, just like logging.

But even limits could easily be subverted. Just *how* easily was demonstrated to me one afternoon in El Rosario. The conference was over; the scientists, bureaucrats, and farmers had made their proposals, most of which had to do with compensating local residents for preserving the forest. Threats had been made ("If you do not give us what we want, we will take what *you* want"). The lepidopterists wondered how their discipline had become a branch of social science. The economists had declared that environmental consciousness must come after rural development, while the preservationists worried that economic development would destroy the environment before any consciousness could take hold.

Bill Calvert, who had heard it all before and did not place

too much stock in talk in any event, was ready to go out and do some real work. The Mazda was resaddled, a new tape popped into the microrecorder, the digital scale recalibrated. Calvert wanted to visit the preserves and get some butterfly weights. He was still guessing that despite their long journey, the butterflies would be fat and in good shape. Two other naturalists, both Americans who had never seen a Mexican overwintering site, accompanied us. It was mid-November, and the monarchs had just recently started to return to the area.

We drove to Ocampo, a small settlement that led to the spot where the trail began, and followed the road farther in, to El Rosario. I had been to this very place just three years before, but I did not recognize it. Where there had been a single dirt track leading to the preserve, there was now a wide concrete thoroughfare; where there had been a couple of shacks from which boys peddled bottles of soda, there were now fifty such structures shingling their way up the path, filled with trinkets, film, and food. There was a designated parking area and small children who greeted us and told us where to park and asked for money to watch our stuff. We had come on opening day—the first day of the season. A party was in progress; we had just missed the parade. Hundreds of paper butterflies were scattered on the ground. But the real action was in the air. We counted 103 monarchs in forty-five seconds.

We paid the fee and started up the trail, accompanied by a Mexican guide. Although the trail was well marked, guides were required. Think security guard here, not interpretive nature guide.

While there were thousands of butterflies hanging in the

air, the most dramatic clusters were about fifty feet off the marked trail, in the woods. Between us and the butterflies was a barbed-wire fence, the serious, three-string kind meant to keep people like us at bay. But when one of the American naturalists asked our Mexican guide for the fourth time if it would be possible for us to get a closer look—indicating that he would make it worth the guide's while—the guide looked left and right, then quickly lifted the fence, signaling us to slip under.

The four of us hustled to a secluded spot, sat down at the base of an oyamel, and just looked. Calvert whispered that there were about five million butterflies above us. He apologized to the naturalists, whose eyes could hardly take it all in: "There'll be a lot more in a few weeks," he told them.

Later we hiked out surreptitiously and found ourselves in the middle of a newly reforested part of the preserve, where we tried with moderate success to avoid trampling the seedlings. Our guide had pocketed eight dollars for lifting the fence. What incentive had he had not to?

Chapter 3

NINETEEN NINETY-SEVEN WAS a good year for eastern monarchs. All across their range, from Texas to Canada, and especially throughout the Midwest, they were seen in unprecedented numbers. "On 28 August I witnessed perhaps the largest flight of monarchs I've ever seen," a man from Ancaster, Ontario, e-mailed the D-Plex list. "This was south across the extreme western end of Lake Ontario in southern Ontario at Hamilton. . . . In the two hours, from 3 P.M. to 5 P.M., that I observed before the flight stopped, I estimated that about 120,000 passed. It's quite possible that they started at 10 A.M., as conditions were good throughout. If so, more than a half million passed on this single day."

A month later, from Bronte, Texas, Jerita Taylor, a high school science teacher, reported, "Each year we have a good showing of monarchs, but this year was spectacular. Thousands of them stopped over in our area. We all turned on our water sprinklers to help them on their journey. (Citizens are encouraged to water their lawns during the migration in Texas, to provide water for the migrating butterflies.) Many of the children had never seen them before, and the school yard was covered on Monday, September 29, because of a heavy dew that morning. We are not very proficient in counting or estimating the numbers that visited us, but the ones who know say this was a banner year. . . . They covered the ground, shrubs, trees, wherever there was water."

And then, on October 28, came this message from Monterrey, Mexico: "Since last Sunday, the twenty-sixth, we have witnessed what I think is one of the most impressive migrations in years. Just imagine—on the radio and TV, municipal authorities are appealing to the citizenship to slow down their cars on the major routes to reduce butterfly casualties!"

It was the next week that Bill Calvert and I headed to Mexico ourselves, hoping to catch up with the butterflies.

Before that, back home in the mountains of New York State, the numbers had seemed unusually high to me, too. This was just a feeling, since I hadn't been keeping count. In the decade we had lived in our shambling old house at the edge of the wilderness, I had grown so used to looking out the kitchen window to the pond and the fields and mountains behind it that I carried within me a mental image of the landscape. It was not that I could draw it from memory, or even catalog its parts. It was that I knew, with the quickest of glances, when something was askew. The red fox that sometimes skulked around the far field, the deer that browsed at the periphery of the forest, the snapping turtles near the abandoned beaver lodge, whose round black backs took in the midday sun—I would always know they were there before I had fully registered what I was seeing.

The butterflies were like this, too. In May I might glimpse only a shadow as it crossed the driveway, but I would know an eager mourning cloak had unfurled the leafy bedroll where it had spent the winter. After that I'd keep my eyes open. The yellow, black, and blue tiger swallowtails would be next, and then the white admirals, which were really mostly black, then the red admirals, then the great spangled fritillaries; it was as if one species begot the other. Then, finally, in July, the first monarchs would arrive, and I would know because my eyes would have grown accustomed by then to the anxious flight of big orange fritillaries, the way they darted this way and that, and then there would be this other winged creature, big and orange, too, but even-keeled and graceful.

ON JULY 7 of that year I saw my first monarch and followed it to the milkweed patch behind the basketball hoop at the edge of the driveway. The milkweed, which is entwined with a stand of raspberry canes, was doing the line-out-the-door kind of business most restaurateurs only dream of: it was buzzing, wing-to-wing, with hundreds of nectaring bumblebees and butterflies. In the days that followed I would sometimes check the undersides of the milkweed for monarch eggs, and on the twenty-third, at midday, I saw one. It was white and about the size of a piece of sleep in the eye. Three hours later, when I brought my daughter out to see it, it was gone. We inspected the leaf carefully and still almost missed the tiny, striped eyelash of a caterpillar that had just hatched out. We checked other leaves: more caterpillars. The patch was "in production."

It stayed like that for weeks, a bevy of mating and egg laying and egg hatching and caterpillar growth and metamorphosis. Egg, larva, chrysalis, butterfly: it was the big show in its smallest, most commonplace rendering. My notebook entry for August 19 said it all: "Monarchs in all stages everywhere now."

THAT YEAR the fall migration to Mexico had been preceded by a remarkably prolific springtime population. "The remigration in the spring of 1997 was nothing short of spectacular," Chip Taylor noted for Monarch Watch. "The numbers of adults, eggs and larvae reported by observers were truly amazing. The monarchs also appeared to arrive earlier than usual in many locations. This may have been due to favorable weather but the large number of monarchs could also have been a factor."

What happened then was that the butterflies spread out across their entire range, from Texas up to Canada, from the Atlantic to the Rockies. Everywhere people might have seen monarchs, they did. In Lancaster, in Mankato, in Newark, in Buffalo, in Lincoln. And then they began to see them in unlikely places, too, such as Saskatoon, Saskatchewan, two hundred miles north of any known milkweed. There were so many monarchs looking for places to breed and nectar that the range was pushed to its limits. Nonetheless, according to Professor Taylor, "favorable reports continued throughout the summer, leading me to speculate that the fall migration would be extraordinary. It was."

So there were lots of butterflies, and people were thrilled to see them in such large numbers, thrilled to walk out of their houses and find them basking in their driveways or clustered in the pines or resting on the ledges of urban skyscrapers. The number of postings to Monarch Watch increased, too, as observers from all over North America recorded their exclamations. It was exciting to watch the numbers and places appear on the screen—exciting the way an election is if your candidate is winning, exciting like the Women's World Cup.

In the Blue Ridge foothills of Virginia, though, one man was brooding: Lincoln Brower, the former Stone Professor of Biology at Amherst College, the Distinguished Service Professor Emeritus of Zoology at the University of Florida, and the current Research Professor of Zoology at Sweet Briar College. Brower looked at the numbers and was unimpressed. No—it was more than that. Professor Brower was *disturbed* by the numbers of monarchs he was seeing. They scared him. To Brower, so many monarchs were a sign of trouble.

LINCOLN BROWER is a generous, thoughtful man. An academic for forty years, he looks the part: tousled gray hair, glasses, rumpled corduroys, elbow patches on his sports coat. In many ways, Brower is the yang to Bill Calvert's yin. Where Calvert spends most of his time in the field, Brower spends his in the lab. Where Calvert is a loner, Brower has acolytes—dozens of young, loyal graduate students scattered across the world. Where Calvert is rail-thin, Brower is . . . rounder. Where Calvert is laconic and cool, Brower is vocal, and sometimes ardent. Where Calvert studiously avoids institutional affiliations, Brower has been in the thick of academia for most of his life. The university has served him well. Brower is the author of hundreds of scientific papers, many of them the result of original and ground-breaking research. He has been the recipient of a dozen National Science Foundation grants. He holds the Gold Medal for zoology from London's esteemed Linnean Society. It was Brower who convinced the United States government to cosponsor the Morelia conference, Brower who had access to the secretary of the interior. It was he who designed the Mexican butterfly preserves after the 1986 presidential decree. No one, not even Bill Calvert, has a more comprehensive understanding of monarch biology than Lincoln Brower. This is widely recognized. Even the *ejidatarios* had heard of Professor Brower, and if they resented his intrusion into their lives and livelihoods, they still acknowledged his scientific expertise. It was to Brower, in fact, that Dimas Salazaar, the farmer from Zitacuaro, addressed his plea—Brower, that is, who Salazaar believed could talk to the butterflies. This was because more than anyone else, Lincoln Brower spoke for them as well.

Still, the 1986 decree was a keen point of conflict. It had been Brower's idea to create a core zone and a buffer zone; he expected the buffer to protect the core. Instead, the people who relied on the forest saw the buffer zone as the only economically viable place left to them and stepped up logging and farming there. And of course there was the matter of boundaries: who could tell what was buffer and what was not? A certain amount of illegal logging was the result of this confusion.

Over time Lincoln Brower had come to agree with the *ejidatarios:* the system was flawed, and stupid. But where they wanted the decree to be scrapped altogether, he wanted it to be rewritten: the core would be expanded, and there would be no more buffer. This did not endear Brower to the *eji-datarios*. Earlier in the year, at a meeting to discuss buying the land outright, one of them had pulled a gun on him. Brower had been annoyed. The man was getting in the way of a serious discussion.

"It's not just that I want to protect the oyamel trees per se because I'm a tree hugger," Brower explained one afternoon in Mexico, as we drove to Cerro Pelón, one of the Mexican forest preserves that he and a graduate student were mapping by examining satellite data to monitor the density of the tree cover there. We had been driving around all day, looking at trees and collecting samples of nectar. This was for another study Brower was doing, one that would measure the food supplies available to butterflies in the area; he was convinced that if agriculture—especially industrialized agriculture, with its reliance on herbicides—was established too close to the colonies, the consequences would be devastating. "This is a unique ecosystem, and the whole damn system is collapsing," he said.

True though that might be, it was also classic Brower—definitive, encompassing, uncompromising, gloomy. (From a paper published in 1994: "[The nine overwintering sites] harbor virtually the entire gene pool of the eastern population. . . . The small size of the area portends disaster for the future of the eastern population.") Depending on whom you asked, the professor was an Oracle or a Cassandra—something mythic, in either case.

Brower was not by nature a melancholy man. The state of the Mexican forests weighed heavily on him, and it angered him, but it had not caused him to lose hope. At sixty-six, he was far too involved in life to be overwhelmed by it. He had started collecting butterflies when he was five, and though he no longer pinned them, he was still actively pursuing them. Butterflies, particularly monarchs, were the text on which everything he did was written. To say he was consumed by them would only begin to touch on the depth of his commitment and the emotion he brought to his work.

"I've probably handled a hundred thousand monarch butterflies in my lifetime and still see them as magic," Brower told me months after we had left Mexico, when he was showing me around his laboratory at Sweet Briar College. "I think of them as magical bottles of wine: you can pour it all out, and when you go back, it's full again. There is no end to the questions you can ask."

The lab was a mess, with books and papers and pipettes and pinned collections of lepidoptera, one made when Brower was a high school student in New Jersey, piled high on tables. Five red Mylar balloons that said "I Love You" were stuck to the ceiling, their white grosgrain ribbons suspended in the air like jet trails. In one corner of the room were three steel freezers, the size and style of those you might see in a college cafeteria. I opened one. It was stocked, from floor to ceiling, with plastic bags full of dead monarchs, their signature orange and black markings just visible beneath the hoar of frost. There were ten thousand of them, dating back to 1976.

"You never know when they'll come in handy," Brower said, though he didn't say for what. He did explain that he had only recently dismantled his lab at the University of Florida and had the contents shipped to Virginia. He was just getting reorganized. He had not really retired, he said, only relocated. (His new wife, Linda Fink, was a biology professor at Sweet Briar and his sometime collaborator.) The "I Love

You" balloons were for an experiment he would be doing the next day on butterfly navigation.

Until recently Brower had shied away from navigation and orientation, concentrating instead on questions that had to do, in one way or another, with chemical ecology. He was an experimental biologist; he had an abiding belief in the necessity, and the beauty, of the scientific method.

"Science is like a language. You have to have a grammar and you have to have rules. It's a universal language because anyone can do it, no matter what his or her native tongue is. What we get from a lot of amateurs is 'I saw five butterflies down at the baseball park on Sunday at five o'clock.' If they went out there every year at the same time and watched what was going on and then published the data, it would be valuable. If you compare a really carefully done experiment with some half-assed little anecdotal report and give them equal weight, that, to me, is not acceptable science."

Brower was getting agitated. Here he was, a biologist, a professor, and yet also the object of more enmity than he, or his station, deserved. And it went way back, to the discovery of the overwintering sites themselves, when the Canadian biologist Fred Urquhart had published hints as to the overwintering sites' whereabouts in *National Geographic,* and Brower, hoping to do some research there himself, had written to ask for more specific directions. When he was rebuffed, he gave the *National Geographic* article to Bill Calvert, and Calvert figured out where the sites were.

"It was January 1977. We were in the forest trying to measure the temperature above the ground. We had thermometers strung at different heights. It was very cold. One of our guides had lit a fire. By chance, Urquhart was there that day, too. I went up to him and stuck out my hand. He refused it and said, 'How did you get here?' Shortly afterward, Urquhart wrote an article accusing me of smoking out the butterflies. Then scores of people who read it wrote nasty letters to the president of Amherst College, where I was teaching, accusing me of killing butterflies. It was very divisive." Brower was cast as a villain among his natural constituents, the amateur lepidopterists.

Urquhart's followers, who numbered in the hundreds, had been tagging monarchs in Canada and the United States, hoping to learn where the butterflies went in the winter and by which routes. To them Brower was an interloper; to him, they were unsystematic. In his language, that meant they were gnats, bothersome but inconsequential. Gracious though he was, courtly though he could be, Lincoln Brower could also be short with anyone who didn't think the way he did. He wasn't out to make friends. He was after the truth, and when he found it, he wanted to let others know. If the truth demanded action, then surely they would want to hear that.

Truth used to be wings pinned to foamcore, properly named. That was when Lincoln Brower was a boy in western New Jersey, roaming the farmland his family owned, and the surrounding woods. He was a collector, a boy who took pride in finding and identifying and displaying. There was an old German entomologist in the neighborhood who took him collecting in the Great Swamp. When the man died, Brower inherited his business, collecting and selling cocoons. He was a boy, making a couple of hundred dollars a year. His parents had been through the Depression. There was a lesson here: you could make a life and a living doing what you loved to do.

• • •

AS A GRADUATE STUDENT at Yale in the 1950s, Brower found himself raising butterflies for a project that his first wife, Jane, was working on: the first controlled studies of mimicry in animal coloring, looking at bird predation on butterflies. The theory was that monarchs were distasteful, even poisonous, to certain birds because their larvae fed on milkweed. Their unpalatability moreover protected not only them but other, similarly colored butterflies such as viceroys, which were not thought to have their own inherent chemical defense. Birds, which rely on sight more than on smell when seeking prey, would see any one of these other orange-and-black butterflies, assume it was toxic, and leave it alone. ("Sometime just pick up a monarch butterfly and pinch it. It will regurgitate. Put a drop on your tongue and taste it; it's really bitter," Lincoln told me.) So Brower raised viceroys, which look a lot like monarchs but are smaller; and monarchs, which are the largest of the black-and-orange butterflies and whose caterpillars feed on milkweed, which makes the cardiac glycosides that are toxic to certain bird species; and tiger swallowtails, which are yellow and black and nontoxic.

The study was a success. Jane Brower was able to show that monarchs were indeed unpalatable. By feeding them to blue jays under controlled conditions, she demonstrated not only that the birds vomited when they ate monarchs reared on particularly toxic milkweed varieties, but that once they did, they learned not to eat them or butterflies that looked like them. It was a learned response.

"Sure enough," Lincoln said, "the monarch butterflies were toxic, and the jays wouldn't eat them or the viceroys.

But they did eat the palatable butterflies, the swallowtails. But birds that had had no experience with monarch butterflies ate them. And got sick."

That was the expected, and hoped-for, result. But to the Browers' surprise, the experiment revealed something else as well: different species of milkweed contain different concentrations of the noxious chemical cardiac glycoside, which in lower doses may be unpleasant to the birds but does not cause them to throw up or die. "If we hadn't thought the tuberosa milkweed was more toxic than the other varieties and had fed the monarchs the nontoxic kind, the experiment would not have worked, and we might have given up," Brower explained. Instead, it set him on a different path altogether, away from his graduate studies in larval cannibalism, through a study of chemical defense in monarchs, to, ultimately, the development of a procedure called cardenolide fingerprinting, which enables scientists to tell where each butterfly comes from by determining what species of milkweed it fed on as a caterpillar.

But in the 1950s, the discovery that different milkweed varieties had different toxicities led Brower to another observation—that there was a kind of protection within the species itself that operated like Batesian mimicry (wherein nontoxic species colored almost identically to toxic ones are able to use their coloring to deter predators, which can't tell which is which). Monarchs that were not especially unpalatable were being protected by the jays' conditioning through eating more toxic monarchs—monarchs that caused "retching, vomiting, excessive bill-wiping, alternate fluffing and flattening of feathers, erratic movements, head and wing jerking, partial eye closure, and a generally sick appearance." In the annals of natural history, it was ground-breaking research.

Not all birds were equally or even adversely affected by cardiac glycosides, however. In Mexico, Bill Calvert had noticed that black-backed orioles and grosbeaks did not shy away from eating monarchs. Indeed, he estimated that in one colony alone, they were killing about thirty-four thousand butterflies a day—or more than a million in that single season.

Predator and prey—that, of course, is how nature works. But it's not hard to see how a scientist might move from an uninflected observation to something more emphatic and urgent—how Brower could easily assimilate this number, one million, and model what it would be if the forests were thinned and the butterflies had less tree cover, or what it might be if they were simply more exposed and vulnerable to attack. Then it would be not just nature running its course but nature with a tipped hand. Clearly the butterflies were at risk if the forest lost its density. Brower was coming to understand that. From there it was not much of a stretch to go from being a research scientist to being an environmental advocate. In a way, the facts demanded it. If the monarchs were going to be protected, then the forest had to be, as well.

"My whole career, up until we confronted the deforestation of the overwintering sites in Mexico, was pursuing questions purely because they were interesting and because there was a historical basis for the research," Brower said the morning we went to his lab. "Even the question 'How do monarchs survive the winter in Mexico?' is a pretty basic biological question. Right now my research is bouncing back and forth between what I can do to show we can't cut these trees and what interesting biological questions I can address."

Later he said something more telling: "I grew up on a large farm in New Jersey. Every place I collected butterflies as a boy has now been turned into a housing development. I had a wonderful trail through the woods where I'd collect underwing moths. I'd paint the trees with sugar, which would attract the moths. Have you ever seen one? They are gorgeous. That land was all sold off and turned into a golf course. One day not long afterward I was out playing golf and I spotted a few white marks from the sugar trail on trees that had been cut down. That just did it. I never played golf again."

In the years since, Brower has made it a point of pride to testify against land development and developers, to take positions on conservation issues across the country and abroad, to marshal the information he has garnered by doing science to push particular political points of view. The science may be neutral, but the scientist is not.

SO BROWER WAS not pleased, the way everyone else seemed to be, with the 1997 bumper crop of monarchs. Amid all the caviling voices on Monarch Watch and the other monarch migration tracking site, Journey North, his was noticeably flat. When monarchs reached Texas in March, much to the excitement of people who lived in Dallas and Uvalde and Johnson City, Brower cautioned, "That's too early. They'll be in the Midwest in April. I don't think the milkweed will be up yet." His point was that the monarchs had to be supported by habitat. They had to eat, or else they would starve. And that was not all. In all the years he had been doing research, he had noticed that monarchs kept to a regular schedule. In Florida, for instance, "they came the last week of March and the first week of April every year, like clockwork." Butterflies that had jumped the calendar worried him.

As respected as Brower was, his worry was largely disregarded. He was like those market analysts who see the Dow cresting 12,000 and say "Correction, correction"—words that no one wants to hear except people selling short. But in biology, no one was selling short, so Lincoln Brower was simply ignored. "Lincoln was just being Lincoln," more than one of the monarch watchers told me, with a "can we get back to the party now?" kind of impatience. In their estimation he was by nature an alarmist.

To some extent this attitude was justified. Although Professor Brower had been the first research scientist (along with the natural historian and lepidopterist Bob Pyle) to make the distinction between an endangered species, which the monarch was not, and an endangered phenomenon, which the monarch migration appeared to be, and though that distinction was accepted and formed the basis for a lot of other research and advocacy, he had squandered a certain amount of goodwill the year before, in a six-hundred-word *New York Times* op-ed piece. Written with the poet Homero Aridjis and entitled "Twilight of the Monarchs," the essay was published just days after an unseasonable snowstorm in the Neovolcanic Belt. Although winter temperatures in the over-wintering sites typically hover around freezing, it rarely snows there. That was one of the reasons, it was surmised, that the butterflies migrated there in the first place. Freezing temperatures could be deadly. Rain could be deadly. Snow most certainly would be deadly.

Lincoln Brower and his colleagues, in fact, had done the definitive studies on the monarch's (lack of) tolerance for low temperatures. As Brower described it, "I wanted to determine the temperature butterflies freeze at, so I inserted

a thermal probe into their bodies, put them into a vial, put
the vial in solution, and dropped the temperature. When the
water in the butterfly freezes, it releases the heat of crystal-
lization. You know that's the freezing point. I did that for
several hundred and found that the freezing point is minus
eight degrees centigrade. Then I wet the butterflies and did
the same thing. They could only go down to one degree
centigrade before they froze."

What this told him was that if trees were cut down and
the canopy was opened up, and it rained or snowed and the
butterflies got wet, they would "lose their cryoprotection."
Big trees—that is, trees with bigger trunks—are more protective than small trees. In winter they are warmer than the
ambient temperature, and they also hold their warmth even
as the air temperature falls. In summer they are cooler. In
both cases this explained why so many monarchs could be
found clustered on tree trunks.

"That's a complicated, nifty adaptation that's interesting
in its own right," Brower exclaimed after describing it.

Nifty as it was, it wasn't nifty enough to stop the illegal
loggers, who were paid a premium for big, old trees. Their
taking them out meant that the monarchs were losing not
only their heater effect but their overhead cover as well—
what Brower called their cryoprotection. But this—the cryoprotection theory—was speculative. Taking out trees was
not. It produced real, hard cash.

And then it snowed. Up in the highlands in the last days of
1995, on the flanks of the Neovolcanics, the snows fell, and
then they accumulated. This was unusual. More typically it
stayed cold in the mountains and sometimes rained, creating
the right microclimate, with sufficient moisture, for monarchs

to spend the winter without drawing down all their lipid reserves. Snow was disastrous to the monarch colonies, especially colonies sequestered in forests that had been thinned. And this was a big snow.

The reports from Mexico were dire. Butterflies were dying in numbers that were exponential. They were falling off trees, frozen. They were lying on the ground, frozen. Everything Brower had been saying for years about the dangers of cutting the oyamel trees was, sadly, coming to fruition. The forest, which had been the winter home of the monarch for ten thousand years, was suddenly no longer able to sustain it. There were reports in the *Houston Chronicle,* the *Mexico City Times,* Toronto's *Globe and Mail,* the *New York Times.* On the D-Plex list the mood was somber. People shared what bits of information they could glean from Reuters and sources in Mexico. They were like people whose loved ones might have gone down in a crash, waiting for some kind of confirmation. And then it came. Brower and Aridjis published their article in the *New York Times.* The first line said it all: "As many as 30 million monarch butterflies—perhaps 30 percent of the North American monarch population—died after a snowstorm hit their sanctuaries in Mexico on December 30."

Thirty million butterflies. Thirty percent of the North American monarch population. The numbers were staggering.

But then the sun came out. The snow began to melt. And millions of butterflies that had been lying on the ground began to warm up and wake from the dead. Not just millions—tens of millions. An estimated thirty million had fallen and been left for dead when the researchers first hiked in during the storm, but when they returned and the final count was done, it was estimated that the winter storm had actually killed far fewer—about ten million butterflies. It was not an inconsiderable number, to be sure, but it was so much less than thirty million that it seemed, by comparison, negligible. Brower looked like a hysteric, the biologist who cried wolf.

Fifteen months later, here he was again, raising his voice, questioning the health of the fecund and apparently robust spring migration. This time his concern was perceived as a kind of intellectual tic, something he could not help himself from doing, and not a serious biological problem at all.

To Lincoln Brower's studied eye, however, the numbers suggested real trouble in Mexico. As he saw it, the butterflies had left Mexico early, unable to find sufficient water and nectar because land clearing right up to the edge of the pre-serves had eliminated their sources.

"In February they started moving down the mountains to get closer to moisture, but instead of finding a mixed pine zone with a rich understory, they found the land cleared out," Brower speculated. "On hot days they would be com-ing down in droves and there would be no place for them to spread out. With their usual staging ground for the spring migration disrupted, the monarchs simply took off and went much farther than usual.

"It's my hypothesis," he added, "and I can't prove it."

Because of their early departure, the monarchs were able to produce more generations in the north prior to the return trip—hence the unusually high numbers. To Brower, then, those numbers spoke not of health but of a decimated habitat. It had been an uncommonly mild spring, ideal for breeding monarchs. But what if it hadn't been? What if there had been hurricanes and tornadoes and late snowfalls? What would have happened to that early generation of monarchs then?

These were the questions Brower was raising, and nobody, it seemed, wanted to answer them.

"I REJECT the argument that since we have starving people, we can't save the butterflies. I totally reject that. Save the butterflies, and you improve the quality of life for those people. And they're going to starve anyway." More vintage Brower. It was getting on to the end of the year. The massive spring migration had turned into what appeared to be a massive fall migration, and now Brower was back in the Mexican mountains, astride an impatient brown mare, waiting to begin an ascent of Cerro Altamirano with Homero Aridjis, the poet laureate of monarchs and the coauthor, with Brower, of the much-maligned article in the *Times*. Aridjis was also on horseback, as was his wife, Betty. Aridjis had grown up in Contepec, a market town that sits at the bottom of the ten-thousand-foot mountain. As a boy, he would hike with his friends up to a field called Llano de la Mula—the "Plain of the Mule"—to picnic and watch the butterflies, which were so thick they bowed the trees. Later he convinced the president of Mexico to protect Cerro Altamirano as well as four other overwintering sites. It had been protected since 1986 because of his efforts; now he wanted to show it to his friend Brower.

The region had been experiencing a drought, and the trail up the mountain was dry and soft as ash. If dirt could be ephemeral, this dirt was, eroding one step at a time so that it disappeared as soon as you touched it. Dust rose like smoke, so much dust you could draw your name in it on the flank of your horse. The animals were walking slowly, losing their footing as the scree dropped out beneath their hooves and rattled away, small avalanches falling behind them, loosening other rocks, making more dust. There were eleven in our party. The trail was disappearing, but we were leaving our mark.

Then the path leveled out, and we scanned the air, the trees, the ground for evidence that this was the way to the monarchs. We saw none. No butterfly wings, no dead butterflies, no butterflies overhead. The oak forest gave way to oyamel fir, and still nothing. Even the birds were still. Homero was somber. He had come to revisit the past but found instead what he imagined to be the future.

And it wasn't just butterflies, the poet said later, when we were resting at Llano de la Mula, a slightly canted alpine meadow ringed by tall oyamel trees. This place used to be full of coyotes, skunks, and rabbits. Armadillo, too. All we were seeing now was ladybugs—thousands of them, crawling and flying in great clusters. Wrong biomass: no one was interested. Homero got up and walked around, back into the forest, looking for the clusters of butterflies he'd been expecting to find. No butterflies, but evidence everywhere of the fires that had devastated this forest fifteen years before, fires started by people attempting to clear land for agriculture, fires that had happened to get out of control. Since then the local monarch population had been smaller and less reliable, Homero said; it was perhaps a direct result of the fires, as well as of logging.

"I feel very frustrated," he admitted, mounting his horse for the ride down. "Every time I come, there are fewer and fewer trees."

"It is very depressing for Homero to see what has happened year after year," Betty added. In his work she often served as his translator, but this was beyond words.

• • •

THE TRIP DOWN the mountain was no less treacherous than the trip up. The horses slipped. The trail kept crumbling away till there was no trail, just dust and rocks. Homero had gone on ahead; Brower was walking and pointing out trees that had gashes in their trunks the size and shape of ax blades. There were many: it was nearly epidemic.

"These flesh wounds will invite disease," Brower explained, "and eventually the trees will die." When they do, they may be hauled out legally, for though it is illegal to take down living trees in this forest because it is protected by the 1986 presidential decree, there is no such injunction against moving dead or diseased timber.

Later, at about five thousand feet, we heard the sound of a chain saw, distant but distinct. A whine, a pause, a whine again. Wood cracked and a tree crashed through the understory. Half an hour farther down the mountain and there was the tree, cut in thirds and tied to the flanks of five donkeys. The donkeys were being led by two men and a boy. "How much will you get for these?" Homero Aridjis asked them.

"Fifteen pesos per donkey," they said. Two dollars.

Chapter 4

IT WAS DRY UP on Cerro Altamirano, but it was dry in Contepec, too—so much so that the authorities declared a water emergency. Showers were forbidden. Flushing the toilet was not looked upon kindly. There were times of the day when tap water was not available at all. There were fires—small ones, but even so, their smoke signaled what lay ahead if the drought did not end. The forests at the edge of town had become tinder, and everyone feared they would ignite. The overwintering sites were at risk, like everything else, but the drought could not be blamed on illegal logging practices there, or on the presidential decree, or on the butterflies themselves. This was an endemic problem, a national problem, having to do with changes to the land, and with population growth, and with the vagaries of weather. So even as the land parched at ten thousand feet and the watershed diminished, Lincoln Brower's hypothesis, that the butterflies were leaving the overwintering colonies early because they lacked water, remained unprovable.

This was of no consolation to Homero Aridjis, who rode back to town brooding. And neither was this: the absence of monarch butterflies at one of their traditional wintering grounds was not meaningful. No one could say that the butterflies weren't somewhere else on Cerro Altamirano; they often changed locales from year to year. Llano de la Mula might not be the St. Tropez, the Aspen, of 1997. And no one could say that monarchs were on the mountain at all. Some years particular overwintering grounds were just not used, and this might be one of those years. But this, too, was of no consolation to Aridjis, who knew that the forests of his boyhood were changing, had changed.

ELSEWHERE THERE WERE reports of heavy concentrations of monarch butterflies, of air viscous with *Danaus plexippus,* places where you could not breathe with your mouth open. Sierra Chincua, near Angangueo, was one of these.

"How many?" I asked Bill Calvert, who brought me there after the conference in Morelia. We were standing on a rock outcrop watching butterflies stream past like spawning salmon. To me their numbers were incalculable, like snowflakes in a blizzard.

Calvert didn't hesitate. "Fifteen million," he said.

This was our last day together. Bill had dragged his groaning, bucking truck up the steep and rutted jeep trail and stashed it in the woods. We didn't have the necessary permits to be anything more than tourists, but out came the scale and the ruler and the logbook anyway. The butterflies were here, and Bill Calvert was eager to know how they had fared after thousands of miles of wind and predators and rain and pesticides and spotty food supplies.

Calvert knew this place. He had spent the better part of fourteen winters camped in these woods, often living by himself in a tent, carrying out his field studies. No one had spent more time here, and even as he walked us farther off the trail, I knew he knew exactly where we were, as if the dense underbrush were macadam marked clearly with a street sign. We turned left, then right, then went straight, losing vertical feet. We were going someplace, though it all looked alike. There were monarchs overhead, clinging to the oyamel trees, and monarchs on the ground, dead. Then the trees gave way to sky and we were standing on bare rock and the sky was a river of orange and black and it was fine that we could not open our mouths because there was nothing, really, to say. After a while we moved off the rock and back through the woods to a clearing and set up shop. Bill snagged about fifty butterflies; I found a clean piece of paper and made a matrix for recording the information. We got back to work, only vaguely conscious of the foot traffic nearby. Sierra Chincua had only recently been opened to the public and was popular with American visitors. We had seen an Audubon tour group earlier in the day.

I looked up. A woman with a pair of Leica binoculars was standing over us, watching intently as Bill Calvert stuffed a butterfly into a glassine envelope the size of a book of stamps, to be weighed and released.

"Aren't you hurting them?" she asked indignantly.

I was about to explain that the envelope protected the

insects from themselves, keeping them from flapping wildly and using up precious energy supplies or damaging their wings, but Calvert beat me to it.

"No," he said. That was all. He looked up and smiled. I knew that sly turn of mouth. Something was up. "Isn't that a black-backed oriole?" he asked idly.

"Where?" the woman asked anxiously, looking up.

"Over there." Bill pointed. The woman put her binoculars to her eyes and walked off, calling to her friends. As soon as she was gone, we picked ourselves up and moved farther into the understory.

"You try catching this time," Bill said, handing me the net. This was like sending in a rookie pitcher at the end of a 12–0 game. There were so many monarchs fanning the air that some couldn't help but fly right into the net. I took a single swipe and landed three hundred butterflies. I got the win!

"You know we could be arrested for this," Bill Calvert would say now and then as we worked. This was not worry talking, it was gleefulness—two sigmas above normal for a seventeen-year-old in gleefulness.

THE MONARCHS LOOKED GOOD. Only a few were tattered or bird-bitten, only a few were thin. Most were bright orange, with full bellies and minimal wear and tear to their wings. They had flown thousands of miles, but there was no way to tell that from looking at them. They had come through just fine.

Although we didn't say so, we were also looking for tags—tiny dots of paper the size of the circle spit out from a hole punch. They would be stuck to the underside of the hind wing. The tags had a sequence of letters and numbers on them—QS498 or NG304—and some other information as well. The tags asked people who found the butterflies, or sighted them, to report their findings to Monarch Watch, at the University of Kansas, which had been tracking the monarch butterfly migration since 1992 and posting the data on the Internet. We looked, but our looking was reflexive. Of the hundred thousand butterflies that were tagged that year, fewer than two hundred were ever found, and only forty-six of them in Mexico. We were in the midst of fifteen million butterflies. We knew the odds and looked anyway. I had tagged twenty monarch butterflies myself, months before in northern New York, and it was these that I was looking for. I had looked in Austin, and in Ciudad Maíz, and in Tula. I had looked in Morelia and in Jamauve. I would look till I left Mexico.

THE BEST TIME to tag a butterfly is in the morning. The air is cool then, and the sun is just rising, and the butterfly, too cold to sustain flight, is nearly paralyzed. Pick it off a tree limb or off a flower and it will seem docile, nearly tame. (It will also seem stuck there: its feet have a natural Velcro on them; it sounds like tearing when you pick it up, but it's not tearing.) You can hold the monarch in your hand, stick it to your sweater. It will not go anywhere. You can collect a bunch this way.

You will need a set of tags, a piece of paper, and a pencil. The pencil does double duty. Use it first to note the time of day, the date, your name, the place where you are, the number on the tag, the sex of the butterfly. (Males have small but prominent black dots on their hind wings—pheromone sacs—and females do not.) Think of the wing as a piece of stained glass, with the black lines as the leading. The distal cell is the largest pane, and the tag will fit neatly within its boundaries. Peel the tag off its backing and stick it there. (Don't worry if some of the monarch's orange scales come off; it will still be able to fly.) Use the pencil to reinforce the adhesive, rubbing the tag with the eraser end. Make sure it is secure. Put the monarch back on your sweater or on a nearby bush. Repeat the procedure with another butterfly. Watch as the sun rises and warms them up. Watch as they take to the air and disappear, carrying your efforts, and your best wishes, and any dream you've ever had of winning the lottery or the trifecta or the Publishers' Clearinghouse sweepstakes skyward with them.

THE MONARCH I WAS really hoping to find had left my yard on the twenty-sixth of August. My daughter had found it near the end of July feeding on the milkweed near the basketball hoop, having deduced it was there from a pile of caterpillar scat and then carefully turning over leaf after leaf until she located it three plants over. She was thrilled. It was as if her intelligence alone had put it there: she thought it should be there, and there it was! To the extent that there was ownership, this caterpillar was hers. It was only about a week old when we brought it to the back porch, put it in the cage we had made from a cardboard box and an old screen, and named it Junior to distinguish it from some of the others there—Biggie, Itsy, and Bitsy among them.

I told myself I was doing this for my daughter's sake, so she could witness the metamorphosis from caterpillar to butterfly. I could describe it to her, or show her pictures or even a video, but none was in real time. Each condensed the experience to the point where *amazing* and *remarkable* and *awesome* were the only words that seemed appropriate. And while it was all of those things, they all missed the nuanced, constant, incremental, and very-rarely-awesome-in-its-particulars way that that remarkable and amazing transformation was occurring. In any case, it was only partly true that it was for my daughter's sake. *I* wanted to see it. Metamorphosis, like resurrection, is a powerful symbol. But what was happening in the cardboard box was not symbolic at all. It was nature investing symbolism with its power.

"Raising" monarchs, as many schoolchildren know, is a blessedly simple task. Supply the caterpillars with fresh milkweed and water daily and watch them grow. In three weeks an individual monarch caterpillar will increase its weight three thousand times and outgrow its skin over and over again, molting five times. When it unzips its striped cuticle for the fifth time, though, it does not acquire a new skin. Instead it seems to turn inside out altogether, and when it is done there is a pale-green chrysalis studded with five gold dots, and no sign at all of the caterpillar that was there.

On the porch we were seeing this with Junior. On August 6 she crawled to the top of the cardboard box, secreted a gluey white liquid, and anchored herself to it. The glue was liquid silk, spun and deposited by Junior's spinnerets. When it was made, she backed up to it and grabbed on with her anal claspers till she was securely fastened. Then she hung there like a health nut in inversion boots, her body forming a perfect J. This step was critical. A few weeks before, Biggie had fallen to the bottom of the box after his chrysalis was made. When he emerged, full-grown, his wing was crumpled and he was unable to fly. He was the monarch that was eaten by mice in our kitchen.

That night, the night of the chrysalis, my daughter told me a story at bedtime. "This will be a little scary and a little sad," she warned me. "Once there was a little girl who woke up with spots all over her. That's the scary part. Soon she made a J and then died. That's the sad part. But she didn't really die because when she came out of her chrysalis she was a beautiful butterfly."

THAT "DIED BUT didn't really die" part was as good a description as any of what was happening inside the acorn-shaped shell that Junior had made. She had gone into it a caterpillar and she would, if all went well, emerge as something completely different—a butterfly. In between she was neither, her larval self having dissolved into a viscous genetic stew that would reconstitute itself into the constituent parts of a butterfly.

After my daughter fell asleep, I opened a book called *The World of the Monarch Butterfly* and copied this into my notebook:

"The change of form and function affects every part of the insect's being, from its senses to the way it moves and feeds. Buds of tissue in the thorax grow and develop into wings. The larva's leaf-nibbling jaws dissolve and new adult mouth parts grow, later fitting together to make a hollow tube through which the adult butterfly will draw nectar. The long intestine shrinks to match the new diet, and sex organs appear for the first time. Long, delicate antennae develop on the insect's head, and the twelve simple eyes of the caterpillar are replaced by the two huge compound eyes of the adult. All these changes are finely co-ordinated, so none comes too soon or too late."

INSIDE HER GREEN ENVELOPE, this was happening to Junior. I didn't know this for sure, of course, since the chrysalis was opaque, but this was what was supposed to be happening, having happened countless times before. The accountability of nature offers its own path to knowledge. I might never have seen a monarch butterfly emerge from a chrysalis, but I could assume it would. It was knowledge that I could count on, that we all do count on—the background knowledge (the sun will rise, the tides will ebb, the trees will bud) that lets us live our lives so exclusively in the foreground.

On August 25 I had the first direct evidence that the metamorphosis not only was occurring but was almost complete: Junior's chrysalis was no longer green, it was black. Then I shined a light on it and saw that it wasn't black at all, it was transparent. The black I was seeing was part of a wing. I could see it, too.

Junior was reborn at 8:23 A.M. Her wings were stubby, condensed, and her abdomen was enormous. She looked like a mutant. A pair of long, articulated legs that ended in pincers grasped her recently vacated apartment. Then her abdomen started to heave, pumping fluid through her veins. Her wings opened to full size like a pocket umbrella whose button had been pushed.

I brought the box outside and let it sit in the sun. Junior clung to her chrysalis and swayed in the breeze like a piece of clothing on the line. Her proboscis yoyoed in and out. At 9:25, an hour into her new life, Junior spread her wings for the first time. They were a deep, almost red, orange. And it really was a new life, at least as against how I had imagined her living it (heading south, wintering in Mexico, mating, heading north, laying eggs, dying), for when her wings were fully expanded, I saw that she was not a she at all. Junior was a male.

Two hours later, when his wings were fully hardened, I gave him his tag and held him up to the sun. I said a short prayer that was really just a wish and waited for him to take off. It's so easy to impute emotions to wild creatures, and even easier to have personal feelings for them. It was more than twenty-four hundred miles from my yard to the Neo-volcanics, and I really wanted Junior to make it. He pumped his wings a few times and peered over the edge of my palm like a diver contemplating the deep end of the pool. Then, without warning, he took off. His wings flapped confidently and he moved like a finch, undulating through the air. What was I to him? I wondered. Probably a tree, I thought, as he circled around me three times before landing in the grass near the pond, then high-jumping to a low branch of a white pine. There he sat for what seemed to me, who was just sitting, too, to be a long time. I went inside. Fifteen minutes later he was out of sight. Gone.

So it was Junior I was looking for when I scanned a roost

with my binoculars. Junior I was hoping to find every time I pulled a monarch from Bill Calvert's net and measured its wingspan. Junior I knew I had absolutely no chance of finding, yet continued to look for, which was, when I thought about it, the acknowledgment of, the recognition of, belief, which is its own kind of story.

"MONARCHS ARE really spiritual for some people," the Canadian monarch enthusiast Don Davis said to me one day. "I had one lady tell me she kept a dead monarch in her refrigerator because it was some sort of religious symbol." Don was the sort of fellow to whom people told things like that. And chances were, once they had, he would let everyone else know by posting it on D-Plex—short for *Danaus plexippus*—the ongoing Internet conversation about monarchs run by Monarch Watch. Information came in and information went out, and when it did, Don Davis was usually involved in the transaction somehow. Academic papers, popular articles, videos, meetings, companies' using monarch butterflies in their advertising—he kept everyone abreast of all of these. Writers called him to check facts, film crews relied on him to guide them, teachers sought him out to give lectures, newspaper reporters wrote annual features about his butterfly work, calling him Mr. Monarch. His "Odds and Ends from Don Davis" was a regular feature of the D-Plex list, a random assortment of monarch comings and goings, like the social column of a local newspaper. And that was in addition to his other messages, which sometimes numbered four in a single day. He was the scribe, the keeper of the history, the librarian of the monarch community.

Of course he was retired—how else could he devote so much time to this? A retired, white-haired, Canadian gentleman: that was how I pictured him, though all I knew for sure was that he was from Toronto. The mind plays interesting tricks, sorting the evidence and fitting it into a template. People are usually more idiosyncratic than we imagine. The retired, white-haired, elderly gentleman I had come to know over the Internet did not exist. When I finally met him, on a summer's day in Toronto, I saw that Don Davis was a small, bespectacled man in his late forties who wore his unfashionably stiff blue jeans unfashionably rolled at the bottom. For twenty-five years he had been a counselor at a Toronto home for abused and troubled children, where he sometimes spent the night. He was unmarried. He was devoted to his Children's Aid work; the monarchs were just a sideline. Still, when asked to describe who he was, he did not hesitate. "I'm an amateur field naturalist," he said.

Among people who tagged monarch butterflies, though, Don Davis was more than that: he was a celebrity. No one had had more tags recovered in Mexico, eighteen so far. No single person had tagged as many butterflies as Don Davis, either, an estimated twenty thousand since 1985. But as great as that number was, it was nothing, really, compared to the number of monarchs migrating over all those years—something like three billion.

So Don Davis, an unassuming man, was embarrassed that luck had been confused with accomplishment. "Back in 1985 I thought it would be really neat to tag a butterfly in Ontario and have it picked up in Mexico. So I did some tagging and the next year I had one picked up there," he said. That was all. He wanted it to happen, and it happened. It was a fluke that year, a fluke the year after that when two more of his tags were found, and a fluke every other time. Yet it kept happening.

Don Davis held other monarch records as well: the greatest number of monarchs recovered in Mexico in one year (ten in 1991); the monarch that flew the longest distance (2,880 miles from Brighton, Ontario, to Mexico to Austin, Texas, where it was recovered on April 8, 1998, seven months after being tagged); the earliest tagging date for a monarch recovered in Mexico (August 14); and the most monarchs tagged by a single individual in a single year (seven thousand). He was the only monarch tagger to have been included in the *Guinness Book of World Records* (for tagging the monarch that flew the longest distance). Don Davis: the Carl Lewis, the Wilma Rudolph, the Mark Spitz of the monarch world.

Although he didn't hold the record, Don Davis had been tagging monarchs longer than almost everyone else, too— over thirty years. "In 1967 I heard about a fellow named Urquhart who was tagging butterflies," he recalled. "Bird banding takes more skill, but anyone can tag a butterfly. So I wrote to him and asked if I could volunteer. I was seventeen years old." This was before the Mexican sites had been found and before much at all was known about monarchs' winter behavior. There were theories, though—that they hibernated, that they went to the southern United States, that they went to Central America, that they went every which way and mostly died off. Fred Urquhart, a professor of zoology at the University of Toronto and curator of insects at the Royal Ontario Museum, wanted to know the answer. In 1938 he started putting tags on monarchs, handwritten tabs of paper that he fastened with glue. They fell off the first time it rained. Urquhart spent years perfecting his tags, at last settling on an alar tag that required him to rub the scales off the front forewing and glue it there. Despite the lost scales and the size of the tag, it didn't seem to affect the butterflies' ability to fly. Meanwhile, his bosses at the Royal Museum thought he was a little daft.

"Each year the director asked for a statement of what each member of the staff had been doing so that he could submit a report to the government," Urquhart remembered. "One year I suggested that a statement be included concerning the results I had obtained in following the migration of the monarch butterfly, to which he responded in a rather disapproving tone: 'What would the government think of a staff member's spending his time placing pieces of paper on butterflies' wings?' But I continued to carry out my monarch butterfly hobby . . . and so the study continued on a rather small scale year after year."

Inspired by ornithologists, who had long relied on amateurs in their studies of bird migration, Urquhart and his wife, Norah, decided to try to enlist the support of large numbers of volunteers to tag and then track monarch butterflies. In 1952 Norah put out a call in the magazine *Natural History*. Within a few years there were some three thousand "research associates," scattered across Canada and the United States, engaged in the Urquharts' monarch project. Don Davis was one of these. "In terms of Fred's project, you didn't have to be a rocket scientist or a Ph.D. to make a contribution to a significant scientific project," Davis observed. "People felt good about that."

It was brilliant public relations, too, to call the taggers research associates. It gave them an identity and status. They deputized themselves; credentials were unimportant. The scientific enterprise was democratized, or so it seemed. The Urquharts were still in charge—it was their research—but people felt good about helping them, about contributing to science. They knew, as did the Urquharts themselves, that the professor and his wife could not do this work alone if it was to yield meaningful results. "There is a limit to what one can accomplish in a project requiring the marking of migrant monarchs," Fred Urquhart reminded his taggers in a retrospective message written fifty years after he began gluing strips of paper to monarch wings. Even if the work of the research associates was more enthusiastic than it was rigorous, it still pushed the rock of knowledge a little farther up the hill.

Almost as soon as the help of nonscientists was enlisted, the data began to accumulate and take shape. In the first volume of their *Newsletter to Research Associates, Insect Migration Studies* (1964), the Urquharts reported that a monarch tagged in Grafton, Ontario, had been found in Baton Rouge, Louisiana. Another, tagged at Niagara Falls, had been recovered in Muskogee, Oklahoma. The butterflies were moving south, for sure, perhaps along a variety of flyways. Subsequent years offered more information, which in turn raised more questions. Why did a butterfly released in Morgantown, West Virginia, fly to Laurel, Maryland, for instance? And how did a monarch tagged seventy miles east of Toronto on September 25, 1968, end up forty-nine days later in Havana, Cuba, a flight that would have put it over water for one hundred miles?

Two years later the Urquharts had their first recaptures in Mexico, about forty miles northwest of Mexico City. "We have had very interesting letters from the captors in Mexico stating that these tagged butterflies were found among many thousands of monarchs roosting in the same area so that we now have proof that the migration occurs in large numbers in Central Mexico," they told their associates in 1972. "Our task now is to trace the migration further south, possibly to Central America." In fact, those roosts northwest of Mexico City were only a day or two away from the overwintering sites.

"It has been, and continues to be, a most fascinating study," the Urquharts continued. "When you pause to think about it, you realize that when we first started we were not certain that all monarchs migrated and that for those that appeared to do so we had no accurate data informing us of their final flight destination. We suspected that they flew from the northern United States and Canada to Florida and perhaps along the Gulf Coast, there to remain, returning the following spring. We now know that it is much more complex than that. We now know, from definitive data, that the population from the northeastern parts of the United States and Canada actually [flies] across the continent from northeast to southwest, finally arriving in southern Mexico and parts of Central America—a most remarkable flight for what seems to be so frail an insect."

In reality, there were no hard data about southern Mexico and Central America—this was just conjecture—but even so, the story was accumulating, like snow on the ground. Every year the Urquharts would send out their *Insect Migration Studies* and the mystery would unfold a little further. It was like the first draft of a novel doled out at the rate of one page a year. But it wasn't just a solution to the mystery of where monarchs went in the winter that the Urquharts were after. It might have started that way, but over the years Urquhart and his wife had become missionaries, too—not so much on behalf of the butterflies, but for the scientific process itself.

"Is it not most satisfying to be involved in the project that takes us out-of-doors; that frees our minds of the petty annoyances of life; that brings us so close to the marvelous workings of nature, and trying to answer the many little and big problems that nature presents to us?" they rhapsodized in 1970. "How much pleasure it is for our young people to be engrossed in a project such as this instead of the many other unfortunate pastimes that occupy so many of our young people today. Together we share our experiences; and together we tell others about our activities and publish the answers to many problems for our scientific colleagues."

But as time went on, the Urquharts became less generous with their information, not more. They grew proprietary, even paranoid, and when in 1975 an associate of theirs from Mexico City, an American textile engineer named Kenneth Brugger, made a fabulous discovery, they were remarkably unmoved. As Brugger told it, "I was returning late in the day from visiting my girlfriend and suddenly I [was] engulfed in a flock of butterflies thicker than I'd ever seen before. Millions. So many that they were falling and being knocked down to the blacktop and cars were slipping and sliding on them. I had to stop. I had a big Winnebago. No cars were moving because the road was so slippery with butterflies. I worked my way back to Mexico City and called Urquhart. He wasn't too impressed with what [I'd seen].

"A month or two later he wanted me to run his research

in Mexico. He gave me a lot of false information because he got it secondhand. Luckily, as a child my girlfriend had been in that part of Mexico where the butterflies were. She used to bring lunches to her grandfather in the mountains. She used to flop on the back of an old swayback mare and ride up to where he ate lunch.

"We went through a lot of dangerous territory. People threatened to shoot us. They told us that Zapata had hidden some gold up there and they thought we were looking for that. We kept going up higher every day. I wasn't looking in the air, I was looking on the ground. Monarch butterflies were dead on the ground. The higher we went, the more there were. And we got to a place where they were real thick on the ground, dead. Kept going and kept going and then we saw them—twenty-one trees loaded with monarch butterflies.

"I called the professor and told him what we'd found. He didn't know me that well so maybe he didn't believe me—I don't know. He didn't come down till the next year."

Meanwhile, Ken Brugger continued his search, looking for other colonies. It didn't take him long to find them: "I met a Mexican who said he used to go up in the mountains as a child and that the butterflies didn't always go to the same place every year. He took me up. Between us we found four or five different colonies.

"The butterflies would go up in November and do what I called overnighting. They would stay in one place overnight, then go up the mountain maybe a thousand feet or so. The next day we'd go to that tree and they'd be gone. We kept following dead butterflies on the ground and there they would be, a little higher. They would make about six or seven trips up there till they finally got to an area they could live with, that had the right kind of trees and the right altitude."

The Urquharts showed up the next year, in January 1976. Ken Brugger arranged for their accommodations and took them up the mountain. In their newsletter of that year they reported, "Last year we informed you of the fact that we had discovered the overwintering site of the monarch butterfly in Mexico. We also informed you that there would be an article published in the *National Geographic* magazine dealing with this discovery and how, after many years' effort, we finally located it with your assistance." Such is the fate of "research associates," the sherpas of science. Few people remember that they led the way up the mountain, reaching the summit first. Credit for finding the overwintering sites went to Professor Urquhart and his wife, who hadn't yet seen them.

The *National Geographic* article, when it came out in 1976, caused quite a stir. A mystery had been solved, and the pictures were sensational. Lincoln Brower, who had been studying the migratory behavior of western monarchs, contacted the Urquharts to get directions to the Mexican colonies. The article had been purposely vague, ostensibly to keep the public away, but when Brower, who hoped to continue his own monarch research there, inquired, the Urquharts declined to share the exact location. They had gotten there first. It was theirs.

Undeterred, Bill Calvert took the *National Geographic* and a map of Mexico and made a guess about where to go, got in his van, and started looking. Then came his discovery, and that New Year's Day phone call to Brower, and Brower's hasty trip to Mexico, and the unpleasant meeting in the forest when Lincoln Brower held out his hand to Fred Urquhart and the Canadian refused to take it. Brower and Calvert, Ken Brugger said, were the opposing faction. There would be no meeting of the minds. Not then, not ever.

Once he got to the woods, Urquhart knew he was on to something big—bigger even than the discovery of the roosts themselves. "On the morning of January 18, 1976, during one of our visits (to the winter preserve), we were surprised to notice that a two inch thick branch of one of the oyamel trees had broken off, caused by the weight of the mass of butterflies clinging to it, and had fallen to the ground. As a result, the surrounding area was covered several inches deep with thousands of monarch butterflies that were unable to fly due to the low temperature of 34°F.

"While we were examining the quivering mass of butterflies, much to our amazement we found one bearing a white tag. This was indeed a remarkable coincidence since of over a thousand trees laden with monarchs, this particular branch had one of our associates' tagged specimens on it. We eagerly returned to our base and telephoned to the University of Toronto. After considerable difficulty, since making international telephone calls from rural Mexico is complicated, we finally managed to reach our office where the secretary looked up the record. Much to our delight we learned that butterfly PS397 had been tagged by Jim Gilbert, with the help of Dean Boen and Jim Street, at Chaska, Minnesota, at the University of Minnesota's arboretum."

THE TAG FROM Minnesota was not the first one recovered in the winter colony. Ken Brugger had found another the year before, a discovery Urquhart was not interested in publicizing. "It belonged to a kid in Austin," Brugger said. "He had made his own tags. They were big and clunky, but they worked. I told Urquhart that the first tagged butterfly I found was from this boy. I went to visit the boy and his parents when I was in Austin. He was a nice boy, very interested in butterflies. I asked Urquhart if I could bring him along when I went back to the colonies. Urquhart refused. He wouldn't let me." Nor did he report this discovery in the annual newsletter, Brugger said, since the boy was not one of the Urquharts' associates. Then, a year later, Fred Urquhart found the monarch from Minnesota with one of his tags on it. *That* story was big news.

It was one thing to know that there were millions of butterflies clinging to trees on the side of a ten-thousand-foot mountain in Mexico. It was quite another to know that one of those butterflies had come from the northern reaches of the United States, 1,750 miles away. The monarch had been tagged four months before Professor Urquhart found it. Here, at last, was conclusive proof that North American monarch butterflies migrated, and that they spent the winter alive, not dormant, clinging to fir trees, waiting out the cold.

Chapter 5

*P*ROOF, in science, is a dissembling concept. It suggests one thing, the truth, and means something else: conjecture. Granted, the conjecture is based on evidence, but conjecture of any kind is still an approximation, a best guess. So when Fred and Norah Urquhart happened upon the monarch butterfly tagged by Jim Gilbert and his friends, all they knew for sure was that a single monarch had gone from Minnesota to Mexico. About the other millions of butterflies in the air and in the trees and on the ground they could say nothing at all.

Soon enough, however, the evidence began to mount. More tags were found, and each one reinforced the supposition that the monarchs in the oyamel forests of Mexico had come from the United States and Canada. It was a guess, yes, but a guess that seemed less speculative as time went on. While the Gilbert butterfly, and the hundreds of recoveries after it, revealed a consistent pattern of monarchs moving from north to south each fall, that was all they revealed. They did not prove that the butterflies reached the Transvolcanics intentionally, through directed flight. And they did not "prove" that there had been a migration. Migration is a story that seems to be true, that many hope is true (it's heroic, exciting, against all odds), and that indeed may be true. But it may not be true, too.

IF THE URQUHARTS acknowledged this uncertainty, they weren't saying so. "We, as a dedicated group of Research Associates, can take credit . . . not only in following the migration but also in bringing this unique phenomenon to the attention of the public," they wrote in their 1988 *Insect Migration Report*. "So year after year, we delve even deeper and deeper into the life of the monarch butterfly."

Accurate though that was then, it would not be so for much longer. It had been more than thirty years since the Urquharts recruited their first associates, and they were tired. By the early 1990s they had had enough.

"Fred decided that he wanted to slow down a bit," recalled Don Davis, their most avid helper. "They were in their eighties and were still getting twelve thousand pieces of mail a year. I agreed that the project should decline, but I still wanted to tag. They did not support this at the time. It was their project, and they ran it; if they said tagging was over, it was over. So after working together for twenty-seven years or so, our contact basically ended."

Tagging itself, however, did not end. Davis had his own labels printed up, as did a few other research associates, and continued to carry on. While some of these tags were recovered in Mexico, it was a limited and idiosyncratic effort at best. The necessary populism of the Insect Migration Association, which had drawn together thousands of people who had little in common but an interest in the most common of butterflies, was gone. And with it went the spirit of the effort, the basic ecology of people scattered across the map recognizing, through a small insect, their relation to one another, and to the land, and to the elements.

LINCOLN BROWER, TOO, had his own tags printed, not to continue the work of the Urquharts but to improve upon it. Brower was unimpressed by the kind of science the Urquharts had been practicing with their research associates. It was too fuzzy, he thought, and too impressionistic, to be of much value. That was when he was being generous about it. When he was not, which was often, he considered it an "amateurish, self-serving approach to biology that isn't science."

But Brower, more than anyone one else, knew that this approach was what had always distinguished monarch research and had, in fact, advanced it. Perhaps because they were seen in large groups, or perhaps because they moved over a sizable territory, monarch butterflies had caught the attention of amateur naturalists—direct heirs to the English tradition of field studies—for generations. Indeed, as Lincoln Brower noted in a monograph written in 1994, "the story of the monarch butterfly is a result of the combined observations of professional and amateur lepidopterists over more than a century."

It began with Charles Riley, an Englishman who emigrated to the United States in the nineteenth century and served for years as the official entomologist of Missouri. Like Fred Urquhart, Riley relied on field observers, in his case randomly dispersed across the state, to supplement his own observations. Monarchs were of particular interest. Not only did they appear to congregate, they seemed to move in a consistent way across the Midwest. As Brower told it, "The accumulation of anecdotal notes of monarch swarms from the prairie across the Great Lake States to New England, supplemented by frequent newspaper and signal officer reports of swarms passing over Iowa, Kansas, Missouri, and Texas, finally convinced Riley that the monarch indeed performs a bird-like fall migration." And that wasn't all. Riley also proposed that the butterflies' likely destination was the southern timber forests. A century later, when Fred Urquhart picked up Jim Gilbert's butterfly, Riley was shown to be right.

Anecdotal science may not have been good science, may not have been science at all, but it yielded interesting questions, and in some instances it led to crucial answers. But when the Urquharts were getting out of the business, it seemed that a whole tradition was ending. Monarch biology was becoming the province of academics. Lincoln Brower, for one, was working on sophisticated methods of chemical ecology. And while Bill Calvert was off in the woods making field notes, he was no amateur, either. In Mexico he was known as *Doctor* Calvert. Even Robert Michael Pyle, the lepidopterist who, more than anyone, was able to introduce laypeople to the beauty and the natural history of butterflies through his Audubon field guides, though unaffiliated with any university, had the distinction of a doctorate from Yale.

Enter Chip Taylor—Dr. Orley Taylor—of the University

of Kansas, a man who would look like Father Christmas if Father Christmas wore crisp blue oxford shirts and khaki pants and snacked on bee pollen. And if his workshop were crowded with amber jars of aminoacetic acid and hexane and syringes and microscopes and caged butterflies and "Far Side" cartoons ("Ten Reasons to become an entomologist: number three: Only about a billion species to worry about") and a window box crawling with bees, all attended by a dozen elves who looked remarkably like midwestern college kids. Taylor was professor of biology and an expert on "killer" bees. But in the summer of 1992, after nearly twenty years of bee research, he was looking to move in another direction. "I was exploring several options when Brad Williamson, a high school biology teacher from Kansas City, showed up, and we began discussing monarchs," Taylor recalled. "The Urquhart program was fading, and it didn't appear that the Urquharts would ever summarize their data—or share them. We discussed initiating a tagging program and Brad insisted we had to do it differently—develop a broader base and involve students. I was a bit skeptical, but the idea appealed to the educator in me and we initiated a tagging program with our own money and some I had set aside from an income account maintained with the endowment association. I didn't envision Monarch Watch when I started—or becoming a monarch researcher or expert myself." But that was what happened.

ON THE WALL of a narrow office on the seventh floor of Haworth Hall at the University of Kansas was a map of North America that looked like a rendering of long- and short-haul trucking routes. Scores of colored lines connect places like Duluth, Minnesota, and Cedar Rapids, Iowa; Olathe, Kansas, and Gun Barrel, Texas; New Orleans, Louisiana, and El Rosario, Mexico; and Orion, Illinois, and Abilene, Texas. The one line that wasn't there—not in visible ink, at least—joins Toronto, Canada, to Lawrence, Kansas: Fred Urquhart's hometown to Chip Taylor's. This poster, though formally titled "Forty Years of Monarch Recoveries," was also a map of Chip Taylor's life for nearly a decade. It showed just how much of a monarch researcher he had unwittingly become.

In the summer of 1992, though, Taylor was still basically a bee man, and a little too shy to get in touch with the Urquharts. "Their summaries were getting shorter each season," he said, "and nothing new was being generated, and no summaries of the overall data seemed imminent. In fact, when Fred and Norah stopped their program, Fred was widely quoted as saying nothing had been or could be learned from the recoveries in the U.S. The data say otherwise. Fred simply didn't have the insight, or collaborators who did, to sort the data to reveal the patterns therein." Taylor and Williamson issued press releases that appeared in newspapers from Minnesota to Texas, recruiting volunteers. They were interested less in establishing that monarchs from the North ended up in Mexico than in finding the routes and the azimuths by which they got there.

Once the newspaper articles came out, the phones began to ring in Haworth Hall. Most volunteers were schoolteachers looking to use monarchs in their classrooms, but there were unaffiliated individuals as well. Monarch Watch—the name Taylor and Williamson gave to their project—took off in ways that neither man had foreseen. In its first year a few thousand butterflies were tagged; five years later the number approached a hundred thousand. By then actual, real, crucial, and sometimes anecdotal evidence about flyways, flight and weather patterns, and endurance had been added by Taylor's amateur minions to the scientific record.

In this effort, however, unlike most scientific endeavors, the outcome was less critical than the process. Taylor was tired of teaching bright university students who could answer the questions put to them on standardized tests but had no idea how those questions had been formulated in the first place. He wanted a project that was unscripted, that would not merely engage young people but inspire them to think scientifically. "All I'm doing is trying to provide a building block," Taylor said one day as he walked through the West Campus greenhouse checking his milkweed seedlings. There were hundreds of them. "When I was thirteen I decided honeybees were very interesting and I was going to learn everything about them. I badgered my mother to take me to the library and I got out all the books on bees and read them over and over again. I've got passion in my soul, and that's what I'm trying to inspire with Monarch Watch. Passion. That's why I like beekeepers. They're passionate."

Outside the greenhouse monarchs were chasing one another and nectaring on purple thistle and basking in the sun. There were pearl crescents and buckeyes and cloudless sulfurs, too, and cabbage whites and viceroys. As Taylor walked through the field, all manner of insect life buzzed around him, or lit on his shirt, as if he were Saint Francis of the invertebrates. He picked a honeybee off an aster and held it to my ear. I drew away, taking one large step backward. Taylor laughed and turned it over. "It's a male," he said. "No stinger." Two paces later he stopped and knelt by a small, leafy cottonwood and beckoned me over. "See that?" he said. No, I didn't. "Look at the underside of the leaf," he instructed. He turned it over and there was a nearly microscopic pearl—the egg of a viceroy butterfly. This might have been an entomological parlor trick, but I was impressed nonetheless. "How did you know it would be there?" I asked dumbly. "Because when I was twelve I taught myself which butterflies lay eggs on which plants," Taylor said. "Viceroys prefer small, young cottonwoods." "Oh," I said, almost walking into a trap laid by the menacing (I thought) orb weaver spider that was presiding over two dead monarchs suspended from her web, shrouded in white silk like bodies ready for burial.

The greenhouse milkweed was necessary for one of Monarch Watch's sideline—and somewhat controversial—businesses, sending larvae to schools and individuals eager to raise their own caterpillars and watch them transform themselves into butterflies. Controversial because the larvae, which were also being raised in the West Campus greenhouse, were being shipped all over the country to be released into the wild population, where they posed the danger of infusing strange or nonadapted genes into a local population or causing bacterial infections. Although thousands went out the door of Haworth Hall each year, Taylor was convinced that they were not statistically significant ("Nevertheless," he urged members of Monarch Watch, "we should be cautious and under no conditions should we release diseased monarchs into the natural population"). In any case, Taylor understood that dominion, even dominion over a small insect, could be a route to passion: maybe some of the people who raised these monarchs would care enough about their fate to learn more about the Mexican preserves or about pesticide use in the United

States or about the use of transgenic crops. Maybe some of them would become field biologists themselves, or zoologists, or ecologists. Taylor saw himself as a teacher, a mentor, an inspiration. The caterpillars were essential.

"Could there be serious consequences of releasing classroom-reared monarchs in the eastern population?" Taylor asked readers of the *Monarch Watch Newsletter* in 1998. Then he rephrased it another way: "What might it take to have a genetic impact on monarchs?" These were not rhetorical questions. Taylor seemed to have an endless supply of them. Did El Niño affect monarchs? Did nectaring monarchs prefer one sugar concentration to another? Did monarchs compete with other species? Were caterpillars attracted to or repelled by light?

"We have legitimate questions to answer," Chip Taylor said. "That's why I put them out there. The data from an eighth-grader have the potential to be just as good as those of a retired senior citizen. There is absolutely no reason amateurs cannot get these data. I want to show that anyone can become a scientist."

Taylor had tears in his eyes when he said this. He believed it as thoroughly as Lincoln Brower did not, and when he spoke he was like that rare political candidate who speaks with conviction, or a preacher who is full of grace. Monarchs mattered to him. Lepidoptery mattered to him. But education mattered the most.

"Science is a process of learning from your mistakes," he said. "If I get data that're dead wrong, I know I'm onto something. Failure tells you where to go next. Scientists forget how many mistakes they made along the way. They present their results in a refined way that doesn't suggest they screwed up for four years."

• • •

BECAUSE THE QUESTIONS mattered to him as much as, and maybe more than, the answers, Taylor eventually got in touch with the Urquharts. He tried to, that is, sending them articles about Monarch Watch, and annual reports, and posters, and newsletters. He never heard from them. When he requested copies of the Insect Migration Association's reports, he was rebuffed and told they were for members only, not for public consumption. The Urquharts' collegiality, it seemed, went only so far. Taylor was disappointed. He had a hunch that all the information they had collected, once it was put together, would add up to something, though he wasn't sure what. Something about which routes the butterflies took to Mexico. How they knew to follow those routes might come after that.

Taylor was finally able to obtain a complete set of Insect Migration Association reports, supplied to him by a former IMA research associate. He and his students got to work, mapping the course of every single butterfly that had been tagged and recaptured over the thirty years of the Urquharts' project. A two-mile trip, a fifteen-mile trip, a trip of fifteen hundred miles—every one of them was plotted. It was like seeing a Polaroid develop, watching those lines accrue. When it was done it showed two distinct flyways east of the Rockies, one coastal, the other through the plains, two routes so consistent that they suggested to Taylor that "the monarch butterfly has a general geographic sense. If it's blown off course it can reorient itself to get back to Mexico, like a bird that gets blown off course. That's a pretty interesting suggestion. This is the only insect for which we have such data."

Taylor was also intrigued by what the picture *didn't* show. There was a huge hole in the map east of New Orleans, north to St. Louis, east to Virginia. No monarch from that area had yet been recovered in Mexico. That was critical because Taylor "sensed" that latitude, as well as longitude, factored into the monarchs' trip to the Transvolcanics. "The tagged data don't meet up with our expectations regarding latitude," he said, pointing to the map. "The butterflies ought to be turning right, right near that hole. The data suggest it, but they just don't support it." So Taylor was going to force the data's hand. Or rather, his associate Dr. Sandra Perez was. Perez, a dynamic young researcher who had been a postdoc in Taylor's lab, would be flying to Washington, D.C., and Dalton, Georgia, later in the year with about a hundred Kansas butterflies on ice. After spending three days in a mesh tent to acclimatize, the monarchs would be released, and Perez would record which way they flew. Would they behave like Kansas monarchs and continue heading south, or would they behave like Taylor supposed southeastern butterflies should behave, and head more westerly? Perez was guessing they'd go south; Taylor was betting they'd head west. A beer was riding on the outcome, but the experiment was still some months off.

"This whole monarch thing is so weird," Perez said one afternoon in Lawrence, having driven in from Tucson, where she was completing a second postdoc (on ants), to consult with Taylor, with whom she continued to collaborate on monarchs. "No one ever asks me about my dissertation, ever. Do you want to know the title? It's 'The Risk-Sensitive Foraging Behavior of Carpenter Bees.' So then I work in this lab for a few months and boom, everyone is interested in what

I'm doing. You just say the words *monarch butterfly,* and people are interested."

Perez, though, was being modest. She said the words *monarch butterfly* and people listened because she happened to say them to the three million listeners of National Public Radio. It was May 1997, and she was standing in a field in Kansas, releasing butterflies and then running after them with a compass as the reporter ran after her with a microphone. Perez was conducting a clock-shifting experiment. Under Chip Taylor's guidance, she had collected a number of migrating monarchs and kept them in the lab for nearly two weeks, changing the light and dark cycles to confuse them into believing they were in a different time zone— Hawaii's, by my calculation. Perez had two control groups as well. One was kept in the lab without being clock-shifted; the other consisted of migrants captured in the wild and kept outdoors.

The question Perez and Taylor were asking was quite simple: Do migrating monarch butterflies use the sun to guide them to their winter homes? To find out, Perez released the butterflies one at a time and ran after them till they were out of sight, recording the way they were headed. The heading— the direction in which the monarch's body was pointing, even if it was being buffeted sideways or backward by the wind—was key, since routine vanishing bearings had proved deceptive. Captive monarchs, especially those with low body temperatures, were notoriously weak fliers. They tried to go a certain way but were not powerful enough to succeed.

"When you take vanishing bearings, you get false information," Perez said. "It can't accommodate the wind. Basically you're getting wind direction. The body orientation and the vanishing bearings were markedly different, so I started to record the body orientation—the heading—as well, and when I did, some patterns started to show up. The vanishing bearings of the animals in the wild were what we expected, but the vanishing bearings of the butterflies we had cooled down in storage were all over the place. But their headings were all the same. Once I realized this, I started to do orientation studies looking at headings, not at vanishing bearings."

Once Perez began to do this, the results were pretty stunning. It was midafternoon on a sunny day in an open field on the Lawrence campus. She released the control monarchs. They flew in the predicted south-southwesterly direction, the direction of the Mexican overwintering sites. So far, so good. Then she began releasing the clock-shifted butterflies, and one by one they began to head west-northwest. They behaved, in other words, as if it were nine in the morning. To Taylor's question "Do monarch butterflies use the sun to orient themselves?" Perez's data seemed to chorus a resounding "Yes!"

"The week the sun-compass story aired on NPR, a guy I didn't know showed up in my lab in Tucson wearing a suit and tie," Perez said. "He said he was working on some kind of nanoplane for some agency in Washington and he thought this biological information might be applicable. I guess he thought the sun compass was the tip of the iceberg, but in fact as far as I was concerned it was the whole iceberg.

"A lot of people didn't believe we were getting these results, because they weren't able to get them. They said it was impossible. But I didn't know it was impossible, so I did it."

Adrian Wenner was one of these. A professor emeritus of natural history at the University of California, Santa Barbara, Wenner was unconvinced by Sandra Perez's results. To anyone familiar with the monarch world, this was not surprising. Wenner was a professional naysayer, a gadfly and critic, the one person least likely to be impressed by *anyone's* data. This negativity wasn't personal, and it wasn't spurious. Wenner was a thoughtful, courteous man—and he was smart, endowed with the kind of searing intelligence that one hopes not to cross. Wenner took one look at the clock-shifting experiment and began to tick off its flaws. At first he said nothing, though it disturbed him to see the experiment written up in *Nature,* which he had thought had more exacting standards. But when the *Los Angeles Times* picked up the story, that was too close. Wenner began openly to debunk the experiment, arguing that it was not statistically significant, that the statistical analysis was flawed, and that the whole enterprise was biased because Perez "knew" in which direction migrating monarchs were "supposed" to move, knowledge that might have influenced her outcomes.

"It seems to me that we keep 'getting the cart in front of the horse,' " Wenner wrote to Chip Taylor shortly after the *L.A. Times* piece ran, "letting theory rather than evidence drive interpretation." Wenner also took the data and put them through other statistical analyses and came up with nothing: these tests did not show the data to be statistically significant. In other words, if these tests had been substituted for the one Perez had used, the conclusion would have been contrary to hers. This was another complaint of Wenner's: if data were analyzed in three different ways and only one of those tests indicated significance, the scientist was free to discard the results of the two "failed" tests. Anything that did not support the narrative could be ignored.

"Striving to find out what animals really do in nature is a far more noble pursuit than trying to 'prove' that they do what we might wish them to do," Adrian Wenner wrote in a memo dated September 25, 1997, and addressed to "Those Interested in Monarch Butterfly Biology." Although nominally commenting on the sun-compass experiment, Wenner was registering a much larger complaint. He did not accept the conventional wisdom; he did not believe that monarch butterflies migrated. He knew that the eastern population moved southward in the fall. He knew that much of the western population moved toward the coast at around the same time. But he refused to accept that in either case the movement was intentional. Intent, he believed, was a human attribute. So was wanting the story of the monarch butterfly to be more dramatic than it really was.

Wenner's own explanation of the southward movement of eastern monarchs each fall went like this: "In the fall, monarch adults in Canada and the upper Midwest likely receive some environmental trigger (change in photoperiod or seasonal cold snap) and cease egg laying. When the main jet stream moves south out of Canada, high and low pressure cells become carried across extreme southern Canada and later across the U.S. At that time, monarchs need merely rise on thermals during clearing conditions and become carried toward the south out of the region in which they were reared. If they have reached sufficient altitude in their ride on thermals, the north winds can carry some of them considerable distances toward Mexico." The reason they all seemed to end up in the same place in Mexico, Wenner argued, was simple:

Monarchs were found in the overwintering sites because that was where people expected to find them. In other words, they might be in other places as well, but the world was big, and who was looking?

D-PLEX, where this discussion and the one about the sun compass and those about the effect of logging on the Mexican habitat and anything else concerning monarchs took place, was another of Chip Taylor's inventions. There were eighteen messages posted in December 1995, the first month the site was in business. Less than four years later, in September 1999, there were eight hundred. There seemed to be no end to the controversies, the information, and the queries. Chip Taylor stayed in the background as much as possible, letting other personalities dominate, but then he would appear, fielding questions, noting unusual recoveries of tagged butterflies (in Cuba, Ireland, the Bahamas), and refereeing the fights that swept through the group now and then like the flu. ("Here's a classic example of a double standard in the butterfly community," a professional butterfly breeder named Rick Mikula wrote in October 1997: "butterflies transported from Michigan to Texas, which everyone will think is cute because children were involved. However, were these butterflies infected with anything? Who knows? But when a professional butterfly breeder rears butterflies under laboratory conditions, eliminating any sick stock, [he's a] bad [person]. Under the current double standard no one turns their head when a nine-year-old releases what could be the most infected monarch at the roosting site, but [everyone] screams when someone takes their time and does it right." Soon afterward Lincoln Brower weighed in, directing his reply to Chip Taylor but posting it for all to see: "Rick Mikula's e-mail message on the interchanges of monarchs borders on a lack of civility and does not advance intelligent discourse on what is a legitimate debate about the wisdom (or lack thereof) in making artificial transfers of monarch butterflies between different geographic areas in North America." "I truly hope you did not find my response as uncivil as Dr. Brower did," Mikula wrote back. "I did not mean it to be. . . . The question I posed still puzzles me. I am all for kids' releasing butterflies, but it seems the professionals always get a bad rap. But I must say after that blasting from Dr. Brower it will certainly be the last time I respond to a posting on the list." It wasn't.)

The migration—if that was what it was—was tracked in a haphazard but engaging way, with people all over the country reporting when they had seen their first spring monarch, or when the fall monarchs were passing through, and in what numbers. There was an exclamatory feeling, passed like a torch from writer to writer as the monarchs moved north, or west, or south: *First Texas Sighting! First Monarch to Reach Canada! Monarchs Clustering in Pacific Grove! Monarchs Leave El Rosario!* And nobody seemed to tire of it, not even Adrian Wenner, tending his garden in Santa Barbara. "I continue to maintain that we actually know little about the remarkable migration phenomenon," he wrote in September 1997, in a message that challenged, yet again, the sun-compass theory. "In the meantime, all stages of monarch caterpillars continue to ravage the milkweed plants in our yard and the females continue to oviposit."

Aberrations were noted, too, as when someone observed

a monarch butterfly mating with a queen butterfly. Or when, in mid-November 1997, Don Davis wrote that he had "received a report from a relative today that in the northwest end of metropolitan Toronto . . . he observed a monarch, with wings opened, sunning on the south side of his garage. I might be a bit skeptical of this report, except that I know this gentleman well and I know that he knows what a monarch looks like."

"Thanks to all the taggers," Chip Taylor addressed the group two days later. "Your efforts have once again produced some interesting data. And thanks to all those who have been so gracious as to track us down or send us the information on the tagged butterflies they encountered. . . . Some interesting patterns have emerged. One of these patterns has to do with the relationship between direction and distance. Nearly all long distance flights are south or southwest, but shorter flights show greater variation in direction."

LATE FALL AND early winter were always a quiet time on D-Plex. It was as if the participants, like the butterflies, were in a state of creative diapause, conserving their thoughts and attention for the remigration a few months off. The excitement of being part of the long-distance relay race as the monarchs swept south from Canada to Mexico was, for the time being, over. In October 1997 there were some three hundred messages on D-Plex. The next month there were only about a hundred. It was the same in December. Don Davis offered his "Odds and Ends" a few times, and there was the usual chatter: discussions about monarchs in Florida and tips on buying milkweed at Home Depot and a message from someone in Warsaw alerting everyone to a movie on monarchs that would be airing on Polish TV.

From Mexico, however, almost nothing was heard, and almost no one was writing about what was going on there, either. It was as if the butterflies, having reached their winter home, were now safely in their beds, asleep, and not to be disturbed. But that, of course, was not at all what was happening. Although their metabolisms had slowed, and though their reproductive systems were temporarily shut off, the monarchs were not hibernating. They spent a surprising amount of time in the air, playing what looked like a child's game: sunbathing in the trees until a cloud drew a curtain on that warmth, then rushing madly skyward, flapping. The sound of their wings was startling then, like spontaneous clapping. It erupted, and arced, and fell away. Most of the time, though, the monarchs were huddled on tree trunks and branches, one upon the next until the bark was no longer visible. They were waiting: waiting for the days to lengthen, for the temperature to rise, for their biological clocks to start ticking loudly again.

NEAR THE END of December a cold front moved through Mexico, and there was some concern voiced on D-Plex that the butterflies might have been affected. Since so many monarchs were clustered in such a small place, cold weather posed a danger, not only to individual monarchs but to the population at large. Chip Taylor posted a calming message three days before Christmas. Yes, it had been cold, he said, but not to worry: the monarchs were fine. "Mortality can be severe however when snowfall is followed by cold rain and then freezing conditions," he wrote. "Long periods (a week or more) of cold, rainy weather appear to have the strongest impacts on the monarchs. The monarchs can't move under these conditions and many become 'waterlogged' (wetted) and fall to the ground where they usually die (or are eaten by mice). I don't have all the literature available but the most severe mortality attributable to a particular weather event I was able to find occurred in 1981 (and not '83, the time of the last major El Niño). Even though the mortality was extreme, 80 percent of the butterflies survived this event. Perhaps Bill Calvert could provide more information on this and other causes of extreme mortality at the roost areas."

Two weeks later he did. Calvert was back in Mexico, at the Chincua and El Rosario sanctuaries, and the news was less reassuring. "It rained hard on Saturday, January 18th," he reported. "Tuesday was partly cloudy and cold. Wednesday not a cloud was in the sky and the butterflies did perform! The Rosario colony was quite high, still above the Llano de los Canejos. That's about equivalent to the top of the loop, the same level as the very top of the trail system, but over to the left about 200 yards. It's a kilometer-and-a-half walk to the colony. At this colony we found evidence of many butterflies knocked down from their clusters by the weekend storm. To get to the colony at Chincua you walk about three-quarters of a kilometer down from the Mojonera Alta where we found evidence of moderate to heavy mortality due to the mid-December storm. The colony that was in position near the Mojonera Alta had moved. Only a few weakened butterflies were evident among piles of dead ones."

Soon after Bill Calvert wrote this, Chip Taylor and Sandra Perez made the trip to El Rosario themselves. "One observer told me he had seen piles of dead monarchs up to two feet deep near the top of the ridge at Chincua," Taylor wrote, noting that the local guides there were reporting that the numbers of monarchs appeared to be down from the year before. Taylor, however, was unconvinced. Population size was always difficult to assess, he said, especially when butterflies were spread across a large area, as they were that winter. He was sanguine. The monarchs looked good, winter was nearly over, five tagged butterflies had been found already, and he had seen a mating pair overhead, a sure sign of spring and of the remigration to come.

A few days later it seemed that this "all clear" might have been premature. David Marriott, who ran the Monarch Project, a California-based monarch education program, was also in Angangueo, helping a film crew make an IMAX documentary on the butterflies, and he called Chip Taylor in Lawrence; Taylor relayed his observations to D-Plex: "The overnight lows at the top of the mountain appear to be lower than normal due to the lack of overcast on most nights. Patches of ice and frost are common each morning. Some of the monarchs visiting seeps to get moisture late in the day are evidently becoming too cold to return to the trees and many appear to be dying from exposure overnight. . . . Local residents claim that this is the driest year in memory."

I was away when these messages were posted, and so I read them later as a group, archivally, without the distractions of watching them unfold in real time, though my mind wandered to Contepec and the horses, and how they had been unable to get any purchase because the ground was so dry. It seemed only natural to find a note from Betty Aridjis, Homero Aridjis's wife, in this stack of mail. A fire had broken out in the mountains we had climbed, destroying hundreds of acres of forest. "The two areas which burned were on the Contepec, Michoacán, side," Betty said, "precisely the place where monarchs had arrived in November, although they [had] quickly abandoned the mountain as there was a dearth of water in the area due to a severe summer drought, and on the Temascalcingo, Mexico, side, an area behind the Llano de la Mula." It was believed that the blaze had been started by a campfire gone out of control. "Lincoln," Betty Aridjis wrote to Lincoln Brower, though she had addressed her note to no one in particular, "remember that on our visit to the Llano de la Mula we saw the remains of several apparently recent campfires?" I remembered, too.

IT WAS MONTHS before anyone tried to put the pieces together—to make sense of the anecdotal reports, to fit them together. Even Chip Taylor in Lawrence, the man with all the pieces, was having trouble. "What happened to all those fall monarchs that were seen heading toward Mexico?" he asked in the 1997 season summary. "Did they make it? It's hard to tell. . . . Perhaps many of the monarchs didn't make it to Mexico or died shortly after their arrival. I visited El Rosario on 14 November. . . . The number of monarchs in the air and in the trees was spectacular but as a biologist I couldn't take my eyes off the hundreds, perhaps thousands, of dead and dying monarchs already scattered on the forest floor. How strange, I thought, to have the biological drive to fly all the way to Mexico only to die within days of arrival.

"In February, we saw evidence at El Rosario that monarchs caught in the open or on the ground at the end of the day had probably frozen to death. Cold mornings limited the ability of monarchs to fly to sources of water, and water became increasingly difficult to find as the winter progressed. At El Rosario, the lack of water contributed to an unusual redistribution of the monarchs late in the winter. In late February and March, a large portion of the colony moved downhill to a source of moisture and trees on the property of Angangueo, the adjacent *ejido*. . . . This appears to be the first time in memory that the colony resettled on the Angangueo side.

"The condition of the monarchs at the end of winter probably determines their ability to remigrate in the spring. Was this a more stressful winter than normal? We don't know.

"Unlike the spring of 1997, there have been no reports of large numbers of spring monarchs on the move. . . . What does this mean? Are we in for a normal year, a good year, or a bad year? At this point, we don't know."

OBSERVATION DOES NOT always yield a bad guy, even if the narrative demands it. No one knew if drought had caused the butterflies to move downhill. No one knew why they appeared to be in bad shape, or why the fall migration, considered to be the "best" in twenty years, had not led to overwhelming numbers in the preserves. No one knew what had happened to all those butterflies.

Chapter 6

*T*ORPID, that's the way a monarch in winter is often described, and there were times, sitting by the wood stove in my house in the mountains that January, watching the snow fall and the birds peck at the feeder, when I felt like that myself, as if a full day's work would be to stay warm and dry. The snow would come, and then it would disappear, beaten back by an icy rain that kept tugging on the power lines, often taking them down. My family left for a while, and I was on my own, being careful always to have wood in the woodbox and candles and matches and jugs of water ready for the inevitable hours when the lights would flicker and there would be darkness and chill and a quiet that would amplify the dog's breathing till it sounded like the saw of an ocean tide. There was a game my daughter and her school-mates played called Predator and Prey, where the prey was a migrating monarch trying to avoid the long reach of its many predators. It was important, then, to "think like a monarch" in order to survive, and sometimes, stoking the fire and reading by flashlight on the couch at dusk, I would find myself thinking like I thought a monarch might "think," thinking the most elemental thoughts about water and heat, nothing more.

I called Lincoln Brower to get an update on the situation in Mexico, and he mentioned that there had been death threats made against Homero Aridjis, who now had three bodyguards and was thinking of leaving Mexico for a while. I took out a quarto of Homero's poems and read them one evening in the uneven glow of the fire. "I have no fear of death / I have died many times already," began a poem called "Fray Gaspar De Caravajal Remembers the Amazon." "Day after day / like all men I have sailed / toward nowhere / in search of El Dorado / but like them all / I have found only / the extreme glare of extreme passion."

In those quiet, unmolested hours, I was wondering about passion, too—about why it arose and why it went away, and how it was that a small insect, for instance, was able to give people their voice. Was lepidoptery a way of cleaving to the authenticity of childhood, to a world undistracted by pretensions, the way certain passions of the flesh were not so much about loving someone else as about finding and expressing one's true and essential self? "Passion extinguishes the logic in chronologic," I wrote in my notebook on January 12, a day when rain fell unseasonably and there was thunder and lightning and a wicked yellow sky. What I meant was that passion kept one fully in the present, so that time became a series of mutually exclusive "nows." Passion obviated history.

But what about migration? Nothing demands such complete attention to the present moment as survival, which, after all, is what the concerted movement from one geographic area to another is about. Yet success—getting there—rests on instinct, the repository of history. In the Old Testament God tells Moses to lead the Hebrew people out of Egypt and take them to Canaan, the land of milk and honey. It was the first recorded migration, that forty-year trip to bountiful, and as with the monarch migration, none who started the journey completed it. So how had they known where to go? Had they used a sun compass, relied on topographical cues, followed the stars? Had they been lured by the poles? The Bible says, but not exactly. Through Exodus, Leviticus, Numbers, and Joshua, as the Hebrew people move across the desert, they are led and dragged and prompted by the hand of God.

The enigmatic, improbable, long-distance, multigenerational movement of monarch butterflies has some resonance here. Since it makes so little *sense* that bugs, living serial lives, could find Canaan each year, and since science has not yet offered a sufficient explanation for how that happens, why not call it numinous and leave it at that? It wouldn't be wrong— surely it wouldn't be wrong—but the fact is, it would be small. It would fail to account for intention, if there is any, and for genetic memory, if that is there, and for the force as fundamental as blood or sex. The wind comes up, the rain comes down, the clouds cover the radial light. The asters have withered, the goldenrod, too, but the monarch, moving south-southwest, twenty-five, forty, eighty-nine miles a day, sure in its mission to survive and reproduce, adjusts. Adaptation, the engine of evolution, is always on full throttle. The constant, variable, unseen, unpredictable accommodations made by a migrating monarch to get to where it needs to go, and its ability to make them, are as essential to its evolutionary design as the shape of its wings or its unpalatability to most birds.

THIS, IT TURNED OUT when I caught up with him in California, was Paul Cherubini's point exactly: the monarch was a remarkably plastic creature, an opportunist able to deal with, and even exploit, what came its way; opportunism was the ultimate evolutionary adaptation. Cherubini, who sold agricultural pesticides for a living, was a provocateur in the monarch world, the one person who could be counted on to take the incendiary position—take it and let it roll among the other monarch enthusiasts as if it were a firecracker about to explode. Then, like a bad boy standing on the fringes, he looked with delight on their horror and revulsion. Lincoln Brower, who could often be found on-line sparring with him, liked to call Cherubini "the exterminator."

If Brower spoke for the butterflies from a preservationist's point of view, Cherubini spoke for them from a developer's perspective, though he wasn't one. Rather, he was that rare species of naturalist who despised environmentalists. To him they were corrupt, money-grubbing elitists. He reminded me of some seasonal workers I knew who lived off unemployment for half the year but always voted for whichever candidate vowed to weed out welfare cheats. He was an angry white guy, the kind who always felt left out and disrespected, the kind whose anger—if anyone cared to notice—came from sadness, not from spite.

"I grew up right around San Leandro," Cherubini told me the day he and I went on a road trip together from Sacramento to the Bay Area, looking at the unlikely places monarchs had chosen to breed and roost. "I used to catch monarchs in wild fields, and I saw all these industrial parks coming in and crowding out everything, and there was one particular monarch site that only had five hundred monarchs and it got cut down and I said, 'My God, what's going to happen?' And I read books like Paul Ehrlich's *Population Bomb* that said there was going to be world famine by the mid-1970s and I said 'My God, I don't have a future.' And I got depressed, seriously depressed. And my parents were having marital problems and the psychiatrist wanted to interview me to get a sense about what their problems were but then he realized I had problems, too, and he said, 'You're depressed, why do you think you have no future?' and I said, 'Because these scientists have Ph.D.'s,' and he said, 'That's not right, you're paranoid,' and I said, 'I'm not going to come to you anymore, facts are facts, scientists are scientists.'

"Then when I went to U.C. Davis I had a professor who was able to show me the structure of science and ideology, and also show the business side of science. He was able to show me that a large part of the environmental movement was based on business. I mean, Paul Ehrlich was worth many hundreds of thousands of dollars and had his own private airplane, all derived from those environmental books in addition to collecting his salary from Stanford. And suddenly I realized that especially in [places like] Ivy League colleges it's a big-buck industry to say all this stuff. If you have a Ph.D. you can say the world's oceans are dying and you don't have to be accountable. You can get rich and not be accountable."

We were driving down a superhighway then, eight lanes across, jostled by the truck traffic as Cherubini kept his eyes low to the ground, looking for monarchs and milkweed in the median strips. They were there, though not in the concentrations he had hoped to show me, for the ground was dry and brown and the milkweed had died back. Still, monarchs would stray into his line of vision and he'd point them out as if they proved his point: habitat protection not only was unnecessary, it was a sham. "Some books, articles, and Web sites (including the Monarch Watch Web site) often state that California monarch overwintering sites are 'threatened' or 'steadily disappearing' due to real estate development and hence the western monarch migration is an 'endangered phenomenon,' " Cherubini wrote on D-Plex around the time of my visit with him. "I think the evidence . . . from the most heavily urbanized areas of North America refutes that dogmatic interpretation of the situation."

"I don't think there's a limit to development," he said when I asked. It was the obvious question. I didn't think it was the obvious answer.

"Look at Los Angeles," he went on. "Look at an aerial photograph of Santa Monica and the area down to the airport. Every inch of land is taken up except for the golf courses. There's no such thing as a vacant lot. Monarchs just have a ball there. They go into people's yards in the daytime and drink the nectar. And especially when there's a drought, they just love the water sprinklers."

It was a happy thought, all those butterflies flitting through all those oscillating sprinklers like little kids frolicking on a hot summer day. I could see how, if you believed in monarchs' "having a ball," it could be quite compelling. And that was not all, Paul was saying.

"In Santa Barbara there is a huge Chevron oil and gas

refinery and they have a monarch colony right on the property. It's the largest aggregation in California. You can photograph monarchs right on their billowing smokestacks. When I saw that, I realized the whole idea of fumes' being bad for monarchs was not right. People live to be eighty in downtown Los Angeles."

I wasn't sure how to respond to this. People die from asthma in downtown Los Angeles, too. People have more respiratory problems in Mexico City than in Sioux City. Each of these facts said something—but what?

We were headed for a couple of golf courses near the Bay Area, places that Paul Cherubini thought "broke all the rules" about overwintering habitat. To the north of them there were town houses, to the south the Hayward Airport, to the west the bay. The greens were triangulated in between, open and flat, shaded here and there by eucalyptus trees, especially along the fairways. There was no canopy, no understory, and some seaborne wind. It wasn't like El Rosario, that was for sure. It wasn't even like Natural Bridges State Park, down the coast in Santa Cruz, which, though on the beach, was densely wooded.

"To my mind, if you had all this openness at ten thousand five hundred feet in El Rosario, you'd be flooded with monarchs," Paul said as we dodged the golf balls that were zipping left and right.

"But you'd never have this much open space, right?" I asked, gliding like a monarch toward a stand of eucalyptus where I hoped to find sanctuary from predatory Spauldings and Wilsons.

"Well, they've never had the opportunity to develop anything like this," he said, hustling alongside me.

"But would you want to do that?" I asked. "Why would you want to do that?"

"I wouldn't want to lie about it to prevent it from happening," Paul said. "Tell the facts to the people and it's their lives, their property, their decision."

"But there is no habitat in Mexico like this, is there?" I asked. The forests were at ten and eleven thousand feet. They were on the sides of steep mountains. It was not exactly prime golfing terrain.

"No," Paul Cherubini said as we wended through the trees, before making a break for the other side of the fairway. "But they haven't had the opportunity."

WE WENT TO Jack-in-the-Box for lunch. Paul explained how low-fat foods were unhealthy but no one could say that out loud in public because to do so would be to take on a huge part of the economy. He mentioned that there was so little pesticide residue on apples that they didn't need to be washed. He was concerned about nematodes, he said, but there was little to be done about them. He looked normal, but his ideas seemed a little off kilter, tending toward the second gunman/trilateral commission/Vince Foster conspiracy side of things. He talked about a sideline business he was involved in, collecting wild butterflies for commercial butterfly farms. He sold the monarchs for three dollars apiece; the "farmers" then turned around and sold them for ten dollars each. It was basically free money, and it was coming in fast. He was genuinely baffled when people on the D-Plex list took issue with his doing this; he was even hurt. ("I'm just doing what everyone else is doing," he said. "This is what pays for Chip's research." But it wasn't: Chip Taylor was sending out caterpillars to be raised, not butterflies to be released.) Cherubini held ideas like beliefs. They were matters of conviction, of faith, and seemed to come from a deep and personal place. They did not lend themselves easily to the tests of science, but then, as Lincoln Brower often pointed out, Paul Cherubini killed bugs for a living. In his line of work, science was about what did the job best.

He did know a lot about monarchs, though. Cherubini had begun tagging with the Urquharts when he was twelve, and soon after that caught a monarch that had been tagged in Toronto and released in British Columbia. The tagger was another teenage boy. His name was Don Davis.

"As soon as I caught Don Davis's butterfly, I was hooked," Paul said. "I knew it would be a lifelong interest in navigation. I said to myself, 'I'm twelve. I have a head start on people who are starting to look at this when they're thirty.' I tagged a whole bunch for the Urquharts. They used to send Canadian monarchs to Reno and California to be released in September to see where they'd end up. I found four of them on the California coast. They were acting like West Coast monarchs."

This notion, that there were two distinct monarch populations in North America, had long been part of the monarch canon. Tagging data from some of the earliest days of the Insect Migration Association showed eastern migrants moving in a concerted southwesterly direction, with those breeding west of the Continental Divide moving westward in the winter, toward one or another of the two hundred or so coastal colonies that would form each year in the stands of Monterey pine and eucalyptus that fringed the Pacific. But in fact, as Bob Pyle and Adrian Wenner and others would show, western behavior was not nearly so neat. While a large number of monarchs west of the Rockies did escape the dry inland heat each year by flying to the coast, others in the northern quadrant moved southward, down the coast, from Washington State to California in the case of one butterfly tagged by Pyle. Still others, in the southern part of the range, actually appeared to move northward, on cooling Santa Ana winds, according to Wenner's data. And to confuse things totally, some western monarchs near the Rockies appeared to head southeasterly, in the direction of El Rosario and Chincua. While the goal was always the same—to turn off reproduction, wait out the winter, and breed in the spring—the strategy was dynamic. As Bob Pyle would conclude in his book *Chasing Monarchs,* "The old model of the Continental Divide as a kind of Berlin Wall for monarchs is bankrupt."

Paul Cherubini knew this. He was convinced, having watched monarchs in Rocky Mountain National Park "flying all over the place," that "there is no such thing as western monarchs and eastern monarchs." But Cherubini was reckless with what he knew. As a kind of hobby, he would transfer butterflies from one part of the continent to the other, from an inland region to the coast, from North to South, from East to West, from the mountains to the plains, to see what they would do, arguing that these "experiments" would further disprove the East-West dichotomy. Although many lepidopterists, especially Lincoln Brower, worried about spreading diseases from one part of the continent to the other, as well as about mixing up genetic stocks, and were calling for a moratorium on such transfers, Cherubini was unmoved.

Brower's concern dated back to the early 1960s, when he

had developed a way to "fingerprint" monarchs by using an assay test to analyze the cardenolide content of a butterfly's guts. Cardenolide, the chemical that makes milkweed toxic, is found in varying concentrations in different species of milkweed. Brower had perfected a method that allowed him to determine which kind of milkweed a particular butterfly had ingested and, since these milkweeds grew in distinct regions, thereby to identify where each butterfly had come from. The monarchs he studied in the Neovolcanics, for instance, had all fed on milkweed growing only in the North. Those he tested up North in the spring showed evidence of having ranged in the South. A new test being developed by Canadian scientists reiterated these findings. Using hydrogen isotopes found in monarch wings, it could determine where the butterflies found at the Mexican overwintering sites had come from. With both tests, transferring butterflies from one region to another might queer the results, not because the tests wouldn't work but because they *would*.

"Lincoln Brower wrote a big paper on transfers a couple of years ago, about the potential for unknown diseases and for mixing up [the monarchs'] genetics," Cherubini said, defending his position. "Lincoln also says you're going to ruin the flight-record database because you are adding monarchs that wouldn't otherwise be there that may be recorded by field biologists. But my argument is that it's an unlikely mathematical probability that they'd have an encounter with those monarch butterflies."

The article Paul was referring to had been published a few years before in *BioScience* and was coauthored, along with Lincoln Brower, by almost every key monarch researcher in the United States. Chip Taylor's name was there, and Karen Oberhauser's and Bill Calvert's. It carried so many names, it was like a petition. Bob Pyle's was missing, but he, too, subscribed to the idea of a moratorium, presenting his own arguments on D-Plex and at the Lepidopterists' Society and in the pages of the Monarch Project's newsletter. Pyle lived up in Gray's River, Washington, not a place overrun with monarch butterflies, though they were present on occasion. So the mathematical improbability that Paul Cherubini observed worked the other way around, too: the release of monarchs in exotic locales increased the chances of their being sighted where they "did not belong" and mistaken for local inhabitants or naturally occurring transients. Such "strangers" were confusing, and that confusion could be costly, both to the scientific record and to conservation efforts. "In particular," Pyle noted, "northwestern monarch students are just beginning to get a sense of how they really behave up here, and a few thousand, even a few hundred, releases could seriously disrupt our ability to do so. This is the same problem that Professor Kenelm Philip, director of the Alaska Lepidoptera Survey, has with painted lady released in Alaska: rare natural events are obscured by artificial releases, so the opportunity is lost to explore how the natural phenomenon works."

Another natural phenomenon—one that scientists were at no loss to explore—was disease among monarchs, especially a protozoan spore named *Ophryocystis electroscirrha,* which was passed from monarch to monarch during breeding, egg laying, and roosting. Infected larvae suffered developmental defects and often died shortly after birth. Western monarchs appeared to be infected at a much higher rate than monarchs that bred and ranged east of the Rockies—30 percent versus 8 percent. It turned out, too, that the western parasite was much more virulent than the eastern one. It was a natural-born killer. All the more reason, according to Brower and Pyle and Sonia Altizer, then a Minnesota graduate student who had raided Brower's stash of ten thousand frozen monarchs to do historical studies of parasite infestation, to keep eastern and western monarchs apart.

But simply saying that there were distinct western and eastern populations posed a temptation for someone like Paul Cherubini. Even if calling them by those names meant only that they were *geographically,* not genetically, distinct, and even though genetic studies had shown that there was almost no difference between the mitochondrial DNA of a monarch from the West and that of one from the East (a rather surprising outcome, according to Chip Taylor, since vertebrates tended to "have differences at ten times this level, while other insects showed differences in mtDNA even within a population"). But their designation as eastern and western was a challenge to Cherubini because it was an assertion he thought he could "prove" wrong by taking a butterfly from Salinas, say, bringing it to Minnesota, and tracking its movements after that. In his *BioScience* collaboration Lincoln Brower had posited that eastern and western monarch populations faced different factors during the long winter months, and different struggles when migrating and dispersing in the spring. It was possible, he suggested, that they had different biological responses, too—responses that had some as-yet-undiscovered physiological basis. He worried that mixing up monarchs from the East with monarchs from the West would impair scientists' ability to figure out if this was true. It might even make it *un*true.

"Two purposes have been stated for the current round of

butterfly transfers," Brower and his colleagues wrote. " 'To determine how California monarchs behave east of the continental divide' (Cherubini, 1994) and to determine if the direction of migration is 'innate . . . or determined directly by the butterflies from stimuli perceived in the external environment of the release location itself' (Cherubini, 1995). The first question has already been answered by the Urquharts' transfers. Monarchs captured at Muir Beach, California, and released in North Dakota flew south and were recaptured in Nebraska and Kansas (Urquhart and Urquhart, 1974). The second question, unraveling the influences of genetic and environmental factors on monarch orientation and navigation, is more complex. It is not clear how our understanding is to be advanced by haphazard transfers, which lack a carefully designed protocol and are unrelated to any laboratory experiments."

So it was back to that. There were some on the D-Plex list who thought the scientists were ganging up on Paul Cherubini and whipping him around with their degrees, bullying him into a version of science to which he didn't subscribe. The commercial breeders in particular were aggrieved on his behalf, but they tended to lurk at the margins since their activities were even more reviled than Cherubini's own transfer "experiments." They were entomology's lowlifes, people who treated butterflies as—in Bob Pyle's disparaging phrase— "biodegradable balloons." And now Cherubini had joined their ranks, picking off chrysalids wherever he could find them, hatching out the butterflies and selling them fresh to breeders. The back of his truck was crowded with cages, and the cages were crowded with small green envelopes that swayed like lanterns in the breeze. "I swear to God I put about thirty thousand dollars into monarch research, and this is the first time I'm actually getting something to pay for it," he said. "I'm just doing what everybody else is doing."

OK, he wasn't just doing what everyone else was doing, but he was doing a lot of it. He had two toll-free numbers on which people could call in to report sightings of the tags he'd printed up. He had been down to Mexico a few times, and back and forth across the country, and once to Australia. He was genuinely interested in monarch butterflies, interested in what could only be called a scientific way. But he was not really interested in doing science. He refused to write up his findings; he had turned down a serious offer by Professor Adrian Wenner to coauthor a paper on California monarchs. He was not interested in "peer review," even though he kept on putting out his ideas on D-Plex and watching the scientists knock him about. None of it—not Lincoln Brower's hectoring, not Chip Taylor's patient efforts at damage control each time Cherubini posted a message that contained controversial or erroneous information, not Sonia Altizer's refutations or Bob Pyle's passionate pleas—threw him off course. In the drama to which they were all, even Cherubini himself, contributors, Cherubini wanted desperately, earnestly, to play the bad guy. It was the one role in which he could distinguish himself.

"You know what I think? I think maybe they're scared to death about what I'll find out when I do these transfers. I think that maybe they're scared I'll show that California monarchs can get to El Rosario." Cherubini laughed and looked gleeful.

"Actually," he said, lowering his voice, "I already did that. A butterfly that I tagged in California and [that] was shipped to Montana was found in Mexico in 1992. An eleven-year-old girl found it. And no one called to tell me. Whoever controls the logbook didn't bother to let me know. I was wondering if the California monarchs would go to Mexico or fly back to California. They all went to Mexico. After that I found out that if you're close to the Continental Divide they can go either way.

"Nobody knows that Montana monarch came from California. I didn't tell anyone except my close friends. If people knew, they'd say, 'He's done it again. He's threatened the migration,' " Paul Cherubini said, not unhappily.

STILL, NOT ALL genetic mutation was "bad"—or, for that matter, "good." More often than not the natural world remains outside the realm of moral values; questions of good and bad do not come into play. Even so, human behaviors have an inadvertent tendency to spill, like oil pumped with the bilge, beyond our own species. Of the thirteen kinds of butterflies found throughout the Hawaiian Islands, for instance, only two are native. The rest are immigrants whose ancestors arrived on hay bales and host plants imported by people. No one knows precisely how the monarch, one of these, got to the islands, or when, but by the mid–nineteenth century it was resident and common. Even more mysterious was the appearance of a rare genetic variant, a white monarch butterfly, at around the same time. Not an albino—that would be all white—but a monarch with its black markings intact and white where there should have been orange.

"The orange pigment is the end product of some metabolic pathway," Dale Clayton, a biologist at Southwestern

Adventist University in Keene, Texas, told me, trying to explain how a monarch could become white—or at least not become orange. Clayton and his colleague Dan Petr, another Southwestern Adventist biologist, were the authors of the only field guide to Hawaiian butterflies and had probably seen more whites in the wild than anyone—and even that wasn't many.

"I once chased a white monarch down the street in Honolulu. People must have thought I was crazy, running in and out of traffic," Dan mentioned when we talked by phone, and the image of him dodging cars and trucks to get a glimpse of this exceptional creature made me want to see it, too. "If you can be at the road to South Point on the Big Island on January thirtieth early in the morning," Petr offered, "we'll try to find some whites."

"METABOLIC PATHWAYS MAY have three or four or seven or ten or some number of intermediaries, and it takes specific enzymes to convert to the next intermediary," Dale was saying. It was eight-fifteen in the morning on the penultimate day of January and we were sitting in the pair's rental car, driving the South Point road at something under fifteen miles an hour. "If you have a genetic mutation, then that pathway doesn't go to orange." Dale did some calculations on a piece of paper and handed it over the front seat. I looked at the inscrutable symbols he had written down. It was the recipe for white monarch.

Dale also passed the road atlas. "This is where we are and this is where we are going," he said, drawing his finger along the road to South Point. The two places were essentially the same. "The road is twelve miles long," Dan explained. "If we take it slow we'll have a pretty good chance of finding a white monarch." By then we had started to cruise the shoulder and I was finding it difficult to listen and look at the same time. Suddenly, without warning, Dan hit the brakes and we stopped short and Dale bailed out of the car, catapulted over a barbed-wire fence, and loped across an uncultivated, weedy field. All this happened without their exchanging more than three words: "Balloon plant," Dan said. "Right," Dale said, already pushing past the car door.

Balloon plant was a kind of milkweed common to Hawaii, and as Dale waded through it, a handful of monarchs rose up like dust underfoot. No whites, though. "This is like the fishing business," Dan said to me as we watched Dale swipe his net a few times, then turn and trot back to the car. "Sometimes you catch 'em and sometimes you don't." This was one of those "don't" times.

In another way, it was not slim pickings at all. As we moved slowly toward South Point, the southernmost tip of the United States, the biologists pointed out a gulf fritillary feeding on a passion vine, dozens of tiny bean butterflies, and the occasional banana skipper. Left and right, all along the road, there were doves and skylarks and grazing cows that spread across the scrubby, tree-bent pastureland as if this were the heartland, not the tropics. It was impossible not to sense the ocean, though, and to expect it, and there it was, finally, at the end of a rutted dirt track, the whole wide expanse of it. We stood there straining our eyes—not at the fishermen casting for pompano, but at the horizon and its promise, seventy-five hundred miles hence, of the next continental landfall: Antarctica.

Back in the car Dan vowed to drive even more slowly and to get out and walk more, for he and Dale were convinced that if we were going to see a white monarch, it was going to be here, on this road. It took us nearly an hour to travel eight miles, and during that time, Dale resumed his genetics lesson and Dan offered an abbreviated history of Hawaiian flora and fauna, which began with the arrival, two thousand years ago, of the Polynesians, bearing twenty-three kinds of plants; touched on the fact that of the forty-five thousand species of mammals in the world, only one, a bat, was native to Hawaii; and ended with the observation that there was little to be found in these islands that was indigenous, an observation that held true for the islands' human populations as well. Although we were progressing, it felt as if we were going nowhere, and then, a little desperately, one of us suggested that we stop back at the field with the balloon plant and make a more thorough inventory. Our last chance. But before we could get there, Dan stopped the car short again and pointed to a different place, a hilly field that was home to a herd of cattle. "See the balloon plant?" he said, and this time we all spilled out of the car, nets drawn, like cops on the heels of a wily suspect.

"There!" Dale called. "There!" I saw a cloud of orange monarchs thirty feet away. Dale moved closer, walking on his toes. And then I saw it, too, a single white monarch needling in among and around the others. It was gorgeous, the way it pulled in the sunlight and sent it out again like a high beam. I followed it with my eyes and got dizzy. Dale, meanwhile, was moving with quiet dispatch through the knee-high grass. "Come out, come out, baby," he called, and the white monarch heard and buzzed his head. Dale parried his net like a lacrosse stick—once, twice, three times. "Got it!" he cried as Dan and I rushed up, congratulating him the whole way, and he pulled it out so we could admire it, which we did, again and again.

ONCE I HAD SEEN a white monarch aloft against the blue sky, I let the Hawaiian waters draw me in, sailing in a fifty-foot catamaran up the Kona coast toward Puako. The rugged beach there is often home to the Hawaiian green turtle, which feeds on algae growing in its shallow pools. It is not unusual to find the turtles asleep on the lava outcroppings or dug into the sand, and to mistake their profile for landscape. It was not possible, however, to mistake the gray whales off our starboard for anything. Migrants, they had come from Arctic waters to breed. "The juveniles gain about three thousand pounds in the three months they're here," said the boat's captain. "The babies are said to grow by something like seven pounds an hour."

Back on shore I went up to Puako to get a bearing on where I'd been. It was a lazy expedition, no agenda, and so I sat on the veranda of the Puako General Store eating lunch, aimlessly regarding a pair of cardinals and the occasional cabbage white butterfly—both North American imports, like me. Although I should have been trained by then, it took a while for my eyes to see that there was a hedge of crown flower milkweed not twenty feet in front of me, and monarchs nectaring on its blossoms. Entranced, I moved off the porch to get a better look. The trees were alive with butterflies. Continuing down the road, I spied more monarchs, and more crown flowers, and more monarchs. It occurred to me that by now I must have seen thousands of monarch butter-flies, and still they pulled me down the street as if they had my hand firmly in the grip of theirs. I stopped in front of a small gray house whose entire front lawn had been given over to a flower garden. The garden was teeming with monarchs, and one of them, I noticed, was white. I followed it around to the backyard, where I met the woman of the house, an eighty-seven-year-old native Hawaiian whose face and hands were as topographical as the carefully placed coral that studded her horseshoe beachfront. She was neither startled by my sudden appearance there nor unwelcoming. She pointed to the metal chair beside her on the veranda.

"Don't you have these at home?" she asked as I took a seat beside her and we watched the white monarch chase an orange out to sea, and seemed pleased for herself and sorry for me when I told her we did not. A gust of wind came off the ocean, lobbing the white monarch shoreward, and it lay for a long time in the grass.

The woman told me bits of her story—worked at the flower shop at the Mauna Kea Beach Hotel, used to live up in Volcanoes, raised potatoes, was a retired lei maker. She wanted to feed me, to give me things, to show me pictures, to explain things about natural history. "The monarchs like crown flower because it is syrupy," she said. "Doesn't matter if caterpillars eat my plants. They need food, too." She wanted to know about me, too, who just wandered into her yard the way a butterfly might, blown in with the wind.

And then there was nothing to say, so we sat there watch-ing the white monarch going nowhere and the ocean chas-ing some surfers back to the beach. What is passion? I asked myself again. It is the collapse of space between two or more bodies, I decided as the woman's face drew close to my own. It is strangers meeting in trust because, though their physical histories are unknown to each other, they are connected by what moves them. It is a cliché to talk about love that binds, but love does bind, and that is why passion, especially passion for a thing, is a way of knowing that comes before epistemology.

"My nephew will continue to feed the insects when I am gone," the woman said at last, answering a question that had not been uttered. No; answering a question that had not been uttered out loud.

Chapter 7

FEBRUARY CAME, and March, and instead of slinking away, winter socked into the mountains like a thick fog. It snowed on March 14, a heavy, wet snow that clotted the roads, making them impassable. When the sun came out we put on skis greased with blue wax and flew through the woods as if those skis were wings. Up one hill, down another, stomach to follow, while the chickadees and goldfinches, their feathers mottled as if, having dressed for one season, they were now deciding better of it, moved overhead, no more graceful than we were.

The air was still frigid, topping off at freezing most days, and thoughts of the tropics and of tropical butterflies were overwhelmed by the drifts of snow hugging the kitchen windows and the prowl of sanding trucks moving slowly across the frozen pavement. Ice still lay across our pond, a big, unbreachable pane of it, and on sunny days it was possible to look through it like glass and see last summer's weeds and misthrown tennis balls and other lost treasures.

It was disconcerting to tune in to Monarch Watch and hear Chip Taylor exclaim about spring and the impending breakup of the Mexican colonies. From where I sat, it was neither warm enough nor green enough to imagine butterflies' being able to head north. If there was a human equivalent of the switch that turned off diapause, enabling one's imagination to range beyond the present, mine was disengaged.

Throughout the winter, as the butterflies huddled under the canopy of oyamel trees and the traffic on D-Plex slowed, I kept track of another migrant crisscrossing the Mexican border with the regularity of a commuter. By my count, Bill Calvert had been back to Mexico four times since I said good-bye to him in Contepec the previous November. Now here he was again, as the snow was flying in the Adirondacks, writing from Mexico that the spring exodus from El Rosario had begun.

"The monarchs again put on a sterling performance last week!" he reported on March 16. "During our last day in the area, we witnessed a massive flow of tens of thousands of butterflies flying out of the colony down across a pasture. All of the butterflies within three meters of the ground were flying in the same direction, giving the impression of a massive sheet of orange-and-black-colored creatures streaming slowly downward. At the middle of the pasture there was a seep of water. Thousands of monarchs were at the seep drinking water from the water-soaked mud and from open pools, but the majority were flying on past the seep. Above three meters fewer butterflies were flying in the opposite direction back toward the colony.

"This colony was the Rosario 'bud colony.' It was the lower part of the Rosario colony that had budded off from the main colony, which occupied a site in the same drainage, but at a higher elevation. Each day more butterflies left the upper (original) colony and joined the bud colony. At this point in time (March 11), it was hard to tell whether the bud colony or the main colony had more butterflies. The bud colony was stopped from descending to even lower elevations by fields and pastures that came all the way up to three thousand meters' elevation. Monarch colonies always descend the mountain during the course of the [winter], accelerating during late February and March when the combination of intense sunshine and lack of clouds and moisture in the air warms up the ambient considerably. The descent is almost always associated with a particular arroyo or drainage.

"At Rosario they usually follow the drainage called Arroyo Los Conejos. However, this year they used another drainage about 1.5 kilometers to the northwest of Los Conejos, called the Rio Grande by the locals. When we were there (well into the dry season), there was only a little water flowing in it.

"The lack of forest at the field edge did not stop them entirely, however. During the day butterflies poured out of the bud colony and over the ridge at the little community, La Salud, toward Angangueo. These butterflies are undoubtedly part of the return migration to the United States and Canada. Each day tens of thousands pass through the town of Angangueo. They are all going in the same direction—northward. Back toward Rosario, many thousands are taking nectar from flowering plants, especially eupatorium and senecio along the road to Angangueo. Many of these do not return to the bud colony mentioned above. Instead they bud again, forming smaller aggregations in remnant pieces of woods along the Angangueo-Rosario road. These small remnants of woodland may be very important to them in offering nighttime shelter from cold and predators."

CALVERT'S MESSAGE APPEARED not on D-Plex but on Journey North, an educational Internet site dedicated to tracking the northward progress of a number of spring migrants, monarch butterflies included. Every week from the end of winter to the beginning of summer, migration updates were posted and maps drawn. Texas, Louisiana, Missouri, Kansas, Illinois, Ohio: it was like a national wave cheer as the monarchs swept through.

The butterflies were dispersing across a wide swath of the country. This was one of the main characteristics of the spring migration. Having come from that wide swath and then funneled into Mexico in the fall, the butterflies simply went out of it in reverse, leaving together through the narrow channel of the funnel, then scattering into the big wide world. We tend to think of a migration as a movement from one place to another—from Ontario, say, to Michoacán. The spring migration, from a few very concentrated places in the south to the entire eastern segment of the United States and parts of Canada, seemed less that than a random dispersal. Still, the monarchs were shifting habitats to take advantage of abundant food and places to lay eggs as spring and summer moved north through the flora—the very definition of migration.

Meanwhile, milkweed, though not migratory, was doing a wave cheer of its own, reborn at higher and higher latitudes as the soil heated up, and the air warmed, and daylight limbered, stretching, stretching, and stretching some more.

JUST AS THE fall migration had captured the imagination of scientists and citizens for the better part of a century, maybe more, the spring movement of monarch butterflies had also piqued their curiosity. Once the Urquharts found the Mexican overwintering grounds, they began tagging butterflies there, hoping to recapture them once the colonies had broken up. The assumption was that the butterflies went north; a recaptured monarch would be good evidence to support that hypothesis. But it was one thing to demonstrate that monarchs from the United States and Canada spent the cold months in the Mexican mountains, and quite another to say that the monarch butterflies that had wintered in Mexico spent the spring and summer in North America. It was a third thing altogether to say—to demonstrate—that the butterflies that made the trip south in October were the very same ones that flew north in March. That would mean that those butterflies were living for six and seven months or more. It would mean either that they were recolonizing the entire northern range, moving up the latitudes like a ladder, breeding and laying eggs as they went, or that they were going part of the way, breeding, and then dying. In that case no individual monarch would make the entire round trip. In that case forget bird migration—the comparison would no longer be apt.

How did the spring leg of the migration work? Back in

the nineteenth century, when Charles Riley, the Missouri state entomologist, was pondering the destination of the swarms of monarchs he was seeing each fall, another question was on his mind as well. Riley supposed that the butterflies were heading south, like birds, but then what? Worn, tattered monarchs had been found each spring in the southern United States, but they had merely raised more questions than they answered. Who could know where those butterflies had come from, and when? Even tagging data, when they were later obtained, were inconclusive. How could it be proved that a monarch butterfly tagged in Minnesota in September and found seven months later in Texas, for example, had spent the winter in the Mexican highlands, unless it was recaptured there as well? And the chance of that happening, statistically speaking, was nil.

Enter—again—Lincoln Brower. While his sometime rival Fred Urquhart was occupied with his tagging project, Professor Brower was in his laboratory at Amherst, continuing the work on chemical defense in monarchs that he had begun as a graduate student at Yale. There he had shown that monarch butterflies were distasteful to birds because of the toxicity of the milkweed they ingested as caterpillars. Now he and his colleagues turned that conclusion on its side and examined it from a different perspective. Since the monarchs stored the toxins—the cardenolides—in their bodies, and since different species of milkweed had different and specific concentrations of the cardenolides, Brower and his colleagues surmised that they should be able to determine which plant or plants a butterfly had eaten in its larval stage. And, they reasoned, since the plants were geographically specific, growing exclusively in some places and not in others, knowing which plants it had eaten as a caterpillar would reveal where that butterfly had come from. They called the process cardenolide fingerprinting. It did have that "the jig is up" quality to it.

To test their hypothesis, Brower and his associates collected fall migrants, butterflies at the Mexican overwintering colonies, springtime monarchs from Texas, and monarchs found in the northern United States in June. As they suspected, the first group, the fall migrants, had fed on *Asclepias syriaca,* the big, broad-leaved, common milkweed that grows north of the thirty-fifth parallel. No surprise there. It was the other groups that told them things they could have only guessed before. While both the winter monarchs and the faded ones found in Texas in the spring showed the *syriaca* pattern, those that had been captured in the North, where *syriaca* was prevalent, had the fingerprint of two southern milkweeds, *viridis* and *humistrata.* To Brower this was "definitive evidence" that the successive-brood theory was right: the migration was a kind of relay race in which fall migrants passed the genetic baton in the spring to offspring whose offspring then continued moving north until they had colonized the entire range and it was time to head south again.

The evolutionary adaptation that had led to this kind of sequential migration had another interesting feature as well. *Viridis* and *humistrata* were both high in cardenolides. The monarchs that carried their fingerprints were extremely toxic to birds, while the ones that carried the *syriaca* pattern were less so. While this might seem to put the generation that left Mexico at risk, it did not. As these monarchs reached the northern tier, their predators had not yet fledged their own young; by the time that occurred, that generation of mon-

archs would already have reproduced, and the new generation, born to the southern milkweeds, would take its protection from them.

AT ABOUT the same time that Bill Calvert was writing
from Angangueo that the spring fling had begun, Chip Taylor, sitting in his office at the University of Kansas, was bang-
ing out his own assessment. His was based not on firsthand
observation but on what he had learned, over years of sorting
through the anecdotal information that came his way, of
the monarch's biological clock. This was the same clock that
Lincoln Brower had referred to the previous year when he
worried that the monarchs' departure from the overwintering site two weeks early was a sign not of an overeager butterfly population but of habitat destruction that had served as
an eviction notice.

On Tuesday, March 17, Taylor pointed out that "during
most seasons, the majority of the monarchs leave the overwintering area during the last two weeks of March. Some
pockets of monarchs, perhaps those in the cooler and most
protected sites, remain in the overwintering areas through
the first week of April. Worn and tattered monarchs should
begin appearing in Texas and Louisiana in good numbers in
the next four to eight days, weather permitting." It was a
tense and expectant time for monarch watchers, who were
waiting to see when, and where, the butterflies would land.
It was a time not unlike those few long minutes I remembered from my childhood, when the Apollo astronauts, tum-
bling to earth, would lose radio contact, and no one would
know where they were, we could only imagine them falling,

and imagine the heat, but not really know, and then someone would see it, a streak in the sky, and hear a voice against the silence, and the relative safety of an open parachute and finally the arms of the sea, reaching up, snagging it. Four days, eight days, and until then, where?

Three days after Chip Taylor's message, Gary Ross, a lepidopterist in Baton Rouge, posted one of his own: "First migrating monarch seen yesterday (3/19/98). It was a male [whose] wings were medium worn. . . . No others seen." Another lepidopterist, a young Canadian named Phil Shappert, who was attached to the Stengle–Lost Pines Biological Station near Smithville, Texas, reported finding a worn male a week later. Harlen Aschen, in Victoria, Texas, saw a tattered female around the same time. "No missing parts but seemed exhausted," he wrote. One after the other, reports like these started coming from Texas and Louisiana, just as Chip Taylor had predicted. If this had been mass transit, it would have had an excellent on-time rating.

As regularly as the migratory pendulum seemed to swing, the migration itself was different each year. The numbers were different, the pathways were different, the conditions were different. When the monarchs left Mexico two weeks early in 1997, not only were they early, they were ahead of the milkweed, causing some concern that they'd keep flying north until, not finding any, they died. The 1998 exodus adhered to a more typical calendar, though it was curious how the butterflies abandoned their roosts to move down the mountains to form bud colonies at lower elevations. No doubt this had to do with water, which had to do with drought, which had to do with both weather and logging. But the spring of 1998 brought a related concern as well, one suggested by Betty

Aridjis's anguished message from Contepec that past winter: fire.

"We witnessed many fires burning in forests all over the states of Mexico and Michoacán," Bill Calvert said in mid-March. "Fires were so frequent and dense that a permanent haze was evident in the sky. None of these were 'serious' fires such as the crown fires that we hear about in our northern forests. All were ground fires burning along the forest floors. They created a lot of smoke and, locally, a lot of heat. One such fire was burning near the Chincua colony located in the Arroyo Honda about five kilometers northwest of Angangueo. Although the smoke from this fire was clearly visible, it apparently has not affected the Chincua butterfly colony."

Even so, the fires continued to burn all through the spring, some twelve thousand of them, almost all a result of the unhappy collusion of drought and slash-and-burn agriculture, producing smoke that crossed the border and covered parts of the U.S., too. In Mexico City the air rained soot day after day and residents were asked to stay inside with their doors and windows sealed shut. The monarchs, meanwhile, had little choice but to push through this gauntlet of particulates. Like the wind that carried it, smoke was not itself a predator, but for monarchs heading north it might as well have been one.

FOR THOSE WHO take delight in the sight of a monarch butterfly coursing through the air or dipping into the still of blue asters, the first spring monarch is thrilling. Its now-you-don't-see-it-now-you-do trick, as it appears suddenly and out of nowhere, too small to have been picked up at a distance like the approaching ducks, and unannounced, unlike the jays and redwing blackbirds, can bring you up short. Where I live the first monarch may float in in late May or June, or July at the latest, and claim the black-eyed Susans and goldenrod. Others follow in a desultory way. There may be three in the yard one day, ten the next, then two, then fifteen.

Following the spring migration over the Internet was efficient, the whole map of the country progressively filling up with dots from bottom to top, each dot representing someone's story of a particular moment that connected his or her life to the life of a small winged creature. But following it over the Internet was frustrating, too. The monarchs could seem less real than symbolic, an icon of the natural world and its mysteries rather than the mystery itself. "As you wait for the monarch migration to reach your hometown, survey the area for milkweed plants," Journey North urged readers casting about for something to do. I didn't have to leave the house. There were a good four inches of crusty snow blanketing the ground. I knew there was milkweed under there, but it had not yet awakened.

"We are having very strong south winds, with gusts up to thirty-five mph," Bill Calvert reported from Austin on March 31. "These winds are undoubtedly driving the monarchs north." For some of us, though, not soon enough.

The ice went out of the pond two days later, retreating by the hour till the open water glowed like new skin and salamanders came to the surface, only to be caught by my daughter, who had fast hands. "I'm letting you go to see the whole wide world," she said to one of them, holding it aloft and turning a perfect circle. She had just turned five, and the pond and the ocean of land surrounding our house were her whole wide world.

On April 12 I surveyed the milkweed again. Nothing doing. The next day I flew down to Texas.

I think I expected to see scores of monarchs in Texas, faded and worn, the same ones, generationally speaking, that I had seen in the Adirondacks in September and in Mexico two months later. I also expected to see the butterflies in droves, packs of them nectaring and laying eggs, together for the last time before fanning out and dying. I was wrong about that: they didn't come across the border in swarms and they didn't gather in groups. The spring migrants were solitary fliers, and though their general destination was known, their touchdown spots were scattered and diverse. To someone hoping to see them, their behavior seemed maddeningly random.

"So did you bring the monarchs?" Bill Calvert asked as I slid into the passenger seat of his reliably cluttered pickup, pushing aside a package of pink hot dogs and a book on the philosophy of science as I did. The Bach tape was still on the dashboard, and maps were piled on the floor, and I knew without turning around that his battered net and makeshift extension pole were wedged in the back behind me. It could have been five months earlier, but for the odometer, Calvert's superego, which registered all the miles he had logged going back and forth to Mexico.

"I thought the monarchs were here," I said, "that's why I came."

Bill Calvert rubbed his mustache and looked amused. "Nope," he said.

"Where are they?" I asked. "Farther north?"

Calvert continued to look amused, maybe because we were in the vicinity of a Luby's cafeteria. "I'm remembering now," he drawled, "that you ask good questions." I was remembering, too: his penchant for answering them with a hapless, almost merry "I don't know."

It was hot in Austin. Bill Calvert was in his trademark jeans and scuffed brown oxfords and a T-shirt. The back of his neck was already deeply rouged by the sun. For me, flying south, the seasons had elided spring, and now it was summer. And somehow, though it was only mid-April, I had missed the monarchs. Calvert, who ran the Texas arm of Monarch Watch, said that most of the reports he was seeing placed the bugs in the northern parts of the state. But even there they were scarce.

By then we were off the interstate and on a two-lane county road that made the transition from suburb to farmland to scrub desert in less than half an hour. Cattle were grazing and oil pumps nodded like obsequious servants, but all in all it was pretty empty territory. We turned down a wooded driveway marked Masters School, and Calvert stopped the truck. The place was almost deserted except for a lone rooster hopping around its enclosure, making a racket.

"This is it," Bill said, making a broad sweep with his arms. His arms opened to a wide, grassy, unshaded field. "This," he said, "is my study site."

I looked around. A couple of buckeye butterflies were ambling through the air, and a red admiral clung to a metal fence. Overhead a purple martin flew by, while a pair of yellow sulfurs mated close to the ground. Calvert was right: no monarchs.

He got out the hot dogs and his knife and handed them to me. "Cut these up," he said. "If you don't mind," he added. The hot dogs were slimy and warm, and I told him so. "You didn't think we were going to eat these, did you?" he asked.

"You're the guy who gets his vitamin A at Luby's," I said, slicing the hot dogs to fit into a container that looked like a medicine vial, the kind that pills come in.

"OK, here's the idea," Bill said conspiratorially. "We're going to put the hot dogs in the containers."

I nodded. This much I had figured out.

"Then we are going to lay them around each metal enclosure, inside and out." He pointed to the field. It was dotted with three circular fences, each about fifteen feet across. Inside each enclosure was grass, and milkweed. "We use the hot dogs to see if there are any fire ants here. They're bait— the hot dogs, not the fire ants. I think the ants are major monarch predators." We laid down the vials around each circle as if each one were marking off a piece of pie.

"Now we wait," Bill Calvert said. "I'm hoping that there will be a preponderance of ants on the outside of the enclosure and virtually none inside."

While he waited, Calvert took an inventory of the monarch larvae on the milkweed inside the enclosures, all the while talking into his tape recorder. "Four-inch shoot with nothing on it," he said. "About two-inch with fifth instar larva and fourth instar. Whoa, there's another one. I have an asperula, nine inches, with nothing on it. Sixteen-inch with mature buds."

When he'd finished inspecting the milkweed inside the fences, he turned to milkweed plants nearby but on the other side of the metal, flipping over each leaf to look for caterpillars.

He turned to me. "Well, this experiment is working well," he said. "I only have larvae inside the enclosures." The alarm on his watch began to beep and he bent over to pick up the first vial, which had been stationed outside the enclosure. "Oh, no," he said, "this is no good. There aren't any ants."

Calvert moved on, checking each vial in each enclosure and then the ones on the outside. At the second enclosure he held up the first vial. "No ants!" he called. He picked up the second, third, and fourth ones. "No ants!" he called again. "Only one more to go." He bent down and picked it up and waved it around. It, too, was ant-free. "Now this is what I like to see!" he exclaimed.

"Oh, no, this isn't good," he said a second later. "This is not good." He was kneeling in the third enclosure, frowning as he watched a gang of ants chew on a chunk of hot dog. Two of the five vials had ants in them. "This is not a good result. We're just going to have to nuke it," he said, taking out a foul-smelling poison and spreading it on the ground around the perimeter of the fence. "I hate to use it, but I need an experiment," he said, distracted for a moment as a single monarch, so worn it was almost translucent, came into range. "There's your monarch," he said to me, barely looking up.

The monarch flew around, it seemed, aimlessly. There were bluebonnets in bloom, and evening primrose, but neither seemed to do it for her. She was three feet off the ground and flapping.

"Why doesn't she do something?" I asked Bill Calvert, who looked at me like I was an idiot.

"What I'm saying is," I continued, "if it's the female's job to get through the winter in order to lay eggs, why is she flying around the field and not landing and not laying eggs?"

Bill Calvert, who was on his knees measuring the milkweed plants, looked up and gave me a big smile that, I have to say, made me feel like less of an idiot.

"What I'm saying," he said after a long pause, "is that I don't know."

I TOOK A WALK then and left Bill to his measuring. I watched him from a distance, the scientist at work, crawling around and checking the backs of milkweed leaves for caterpillars. "Got a big one!" he'd call, or "Lots of monarch biomass here!" Calvert was also checking to see which of the two milkweed varieties growing in the field was preferred by monarchs. It was too early to tell, but the results would dovetail nicely with Lincoln Brower's work on cardenolides. If given a choice, would the monarchs choose the milkweed that would give their offspring more protection? One might expect so, and Bill Calvert was eager to see what would happen. A half hour went by, then another. Calvert kept measuring and talking into his tape recorder. "This really gets exciting right around now," he called to me as I went over to the truck and tried to figure out a way to sit in its shade without attracting fire ants into my shoes and pants. "Who else would come and spend three hours in the hot sun except someone who is really excited?" he said.

•   •   •

THAT TRANSLUCENT MONARCH we saw noodling around Bill Calvert's study site was the last one of 1997 that I saw. Two other monarch butterflies settled into the Indian paintbrush that afternoon, but they were bright orange—fresh ones—as were the few I found nectaring along the highway as I headed into Houston the next day. The fresh ones were the freshmen of 1998. By the time I'd start to see monarchs again in the Adirondacks a few months later, they would be three, maybe four generations removed—seniors, or postgraduates, to stretch the metaphor. But that assumed that each generation would do its part. That the chain would not be broken. Yet that, observers in the southern tier began to notice, was not what was happening. There were fewer monarchs, for one thing, or at least fewer monarchs being sighted, as Bill Calvert had told the readers of D-Plex just a few hours before I got off the plane in Austin.

"In general we are not seeing as many adults or eggs and larvae as we did last year at this time," he wrote. "The greatest numbers of adults were seen in late March and very early April. Since then numbers of adults have tapered off. We continue to see eggs, but not very many. It may be a ho-hum year for monarchs!"

But ho-hum it wasn't shaping up to be. The numbers were definitely down. A comparison of sightings reported to Journey North showed that during the first two weeks of April there had been fifty-nine sightings in 1997 and only twenty-three in 1998. And that was not all. Monarchs were showing up in strange places at strange times. One was seen crossing Bancroft Point on Grand Manan Island in New

Brunswick, Canada, on May 4, nearly two weeks before a monarch had ever been seen there before. And then there were the fifty monarchs seen crossing the Gulf of Mexico, flying against the wind. That one caused a lot of speculation, but one man, David Gibo of Toronto, didn't find it peculiar at all. Gibo, as I was soon to find out, was a glider pilot with an encompassing interest in flight vectors. He was also a biologist, a professor at the University of Toronto at Mississauga. Those fifty monarchs had probably been riding thermals, Gibo supposed, and when a thermal that has formed over land drifts over water, the warm air feeding it is cut off and it breaks up—in this case causing the monarchs to descend and forcing them to fight their way back to land. "Were the butterflies simply unlikely individuals that happened to have started too close to the coast when they picked up their first thermal of the day?" Gibo asked. "Probably. Were these sightings unusual? Not at all."

Even so, the numbers were down significantly, and no one knew why. Fire, drought, and logging were the obvious culprits, but so was nature itself. Population fluctuations had been observed for so long that some biologists, most notably Fred Urquhart, thought the monarchs were on a seven-year boom-and-bust schedule that reflected the rise and fall of certain parasitic killers. It wasn't so, at least not the cyclical part, though disease could be rampant. When I got home I called Bill Calvert to get his opinion.

"Why are the numbers down?" I asked him.

"Are they?" he replied.

"Well, the number of reported sightings is down," I said.

"Right," he said, making his point. I decided to change tacks.

"What if the numbers were down this spring? I mean, they were so high last fall, what if the spring population basically crashed? Why would that be?"

There was silence for a moment, and then, if a smile can be audible, I heard one traveling across Bill Calvert's face.

"That's a good question," he said.

Chapter 8

THE MILKWEED BEHIND my house began to reassert itself, slowly at first and then with more vigor. Two inches in the beginning of May; ten more by month's end. Green with promise, it stretched skyward, adding leaves and stem, then more leaves and more stem, like a stunt done with mirrors. The raspberry canes that had lain brown and dormant among the milkweed began to wake, too. Horizontal on the twelfth of May, they had sprung some twenty degrees two weeks later and continued degree by degree, like someone rising from a long and deep sleep, till, heavy with fruit, they stood perfectly vertical amid the tangle of milkweed.

The birds came then, the thrushes and the warblers, and then the butterflies and the bees. Traffic was heavy with all their comings and goings. Pollination, copulation, ovipositing, and predation were undertaken with such diligence that I began to think of the milkweed patch as a small enterprise zone or industrial park. There were beetles and earwigs, ants and wasps. And spiders and grasshoppers. What there wasn't, though, was monarchs. The monarchs were conspicuously absent. In years past there had always been a few in early summer, outliers who had come up from the Midwest on a northern wind, second-generation migrants whose offspring might head up to Canada or stick around and breed the generation that would go down to Mexico. But this was not a typical year. The monarchs were missing, and not just from my backyard, and not just in May, and not just in June. Not one monarch butterfly lit upon the joe-pye weed that grew casually around the perimeter of the property. Not one was sighted on the goldenrod in August. Their absence was a kind of visual silence—an anxious visual silence. The monarchs always seemed about to appear, but then they never did. And no one could say where they were.

But the world, even the world of my backyard, is a big place. It was possible that the monarchs that were not on my property were two miles over, on my neighbor's. It was possible that they were in Franklin County, not Warren County, or in Quebec, not New York. Or in Vermont, or Maine. Just because they could not be seen, it didn't mean they were not there. Empirical knowledge is tricky that way. Things unseen are not necessarily things not there.

Some days, walking around the yard, turning over the newest milkweed shoots to look for monarch eggs and caterpillars, I found it miraculous to think that I had seen monarchs (and eggs and larvae) there in the past. They were so small, and so scattered. But they *had* been there; against the vast backdrop of earth and sky, they had shown up. And if the consolation of empiricism is truth found in regularity, when things stop happening it is unsettling. This was in August. Clusters of mallards and mergansers had already taken to the air. I decided to, too.

"Just so you know," I said to the man on the phone, in the interest of full disclosure, "I really don't like to fly."

"Many people are afraid of flying," he said in a soothing voice. "But they haven't really flown."

"I fly all the time," I said. "I can't tell you how many airplanes I've been in in the past year."

"Did the airplanes have engines?" he asked.

"I assume so," I said.

"So you haven't really flown, either," he concluded.

The man on the other end of the line was David Gibo, the University of Toronto biologist and glider pilot. Gibo had logged hundreds of hours in his Grob 103 ACRO, a two-seat, engineless fiberglass aircraft with a wingspan twice as wide as a Piper Cub is long. A glider is lifted into the air by a tow plane, which unhooks it at two or three or five thousand feet or more, leaving it to find its own way back to the ground. It was in the air that Gibo began to understand how a monarch butterfly could travel thousands of miles and end up on the side of a mountain in Mexico without apparent damage. But it was from the butterflies that he first learned how to fly.

"One September day I glanced out of my office window

and saw a monarch," he recalled. "I had just learned to glide myself and I was looking out the window to see if it was going to be a good weekend to drive out to the glider club. I saw this butterfly coming toward the building and it started going up and got about a third of the way and then stopped flapping. From then on it went back and forth, turning figure eights or circles, I can't remember which, until it went up and over the building. I said 'Hmmn,' and I knew what it was doing, because if it wasn't flapping its wings it was coming down, unless it was in air that was going up. Buildings, like mountains, create lift."

This was in the mid-1970s. Fred Urquhart's *National Geographic* article extolling the seemingly valiant, enigmatic, long-distance journey of the monarch butterfly had just been published. So it was known where the butterflies spent the winter. What was not known was how they got there. And then David Gibo, a wasp guy by trade, saw the monarch catapult the South Building, and it gave him an idea.

"I decided to study flight tactics," Gibo said. "The solution for that seemed obvious. All I had to do was build a powered ultralight aircraft, add a few extra instruments, then fly in the vicinity of the migrants."

But it didn't work that way. Gibo built the plane and hauled it down to Texas at the height of the migration, taking off on a day when he could plainly see thousands of butterflies overhead. See them, that is, till he was part of the sky himself. "All my equipment and theories about air currents could not help me see what amounted to a piece of paper set on edge against the horizon, and a camouflaged one at that," he said. "Look down, no butterflies. Look ahead and to the side, ditto. I finally realized that I could see them if I looked up, but that isn't a very safe way to fly. Especially at low altitudes."

Gibo stopped looking for monarchs when he was in the air and concentrated on becoming an expert flier himself. If he understood aerodynamics better, he reasoned, he would understand how an insect with a three-centimeter-long body could travel forty-five hundred kilometers through the air— a feat, he calculated, akin to a six-foot-tall person's circling the globe eleven times in a row. But the monarch was not a person, it was a bug, a bug with a bug's brain. Or as David Gibo liked to say, it was aerial plankton with a guidance system. How complex could it be?

"Without a doubt, [the monarchs'] annual two-way migrations are among the most amazing accomplishments of insects," Professor Gibo assured readers of his flying, gliding, soaring, and science Web site, Tactics and Vectors, some of whom, he knew, ascribed spiritual dimensions to the unlikely, and seemingly miraculous, journey of the monarch, as if it were nature's vision quest. "Nevertheless, it can't be that difficult. We're talking about an insect. Like all insects, butterflies are strong and resilient, but lack special (i.e., magical) powers and are prone to all the limitations that accompany small body size. Compared to migratory birds, migratory butterflies are much slower [and] have an inferior capacity to regulate their body temperature and an absolutely ridiculous rate of fuel consumption during powered flight. In short, there seems to be nothing to recommend the butterfly body plan, physiology, and nervous system for the task of making regular, long-distance, directed migrations. . . . Nevertheless, each year millions of butterflies, who apparently haven't the good sense to recognize their serious design flaws, somehow manage to make their way across the continent. Apparently, we're overlooking something important here."

What he meant, and he wasn't being pejorative, was that it couldn't be all that complicated. Bugs were bugs. Biologically, physiologically, they were capable of only so much. They were scripted at birth. They followed a distinct genetic set of rules that directed them to fly to the overwintering sites and back. Gibo believed that those rules were discernible. He wanted to uncover them.

"Let's say that there is an international airport and we want to find the regulations that apply to different categories of aircraft," he said to me, reaching for an analogy that he thought I might understand. "We also want to know which runways they take off from, and how it is aligned, and what the requirements for noise reduction are. If I just watched what was happening and recorded it, at the end of a year's observations I would say, 'OK, here are the regulations.' I'd be able to infer the rules."

What Gibo did *not* mean—and this was central to his way of thinking—was that the rules were about probability, the way, say, Adrian Wenner's dispersal model was, or about randomness and chaos, the way a model based solely on weather might be.

"Monarchs don't simply show up wherever the wind blows them," he pointed out. "They seem to show up where it is beneficial to them, at a higher latitude at the right time of the year, where the crops or fruit plants are. And this seems to be more predictable than what we expect from the weather patterns. I think that going and getting descriptive data and exploring them will give more insight into what's happening than using meteorological data and generating hypotheses that way."

• • •

IT WAS THE MIDDLE of August in the summer of the absent monarch butterflies. In less than two hours I would be airborne with David Gibo, flying in a two-seat training glider above the farmland surrounding Arthur, a small agricultural village in southern Ontario, about an hour west of Toronto. We had driven up at midday, stopping for lunch at the local diner, a spare eatery with little to recommend it except that it was there and reliably patronized by members of the York Soaring Association, Professor Gibo's glider club. Though he was so engrossed in a conversation with a fellow pilot that he kept forgetting to look at his menu, I was obsessed with the French fries and chocolate milkshake that I ordered with abandon on the theory that this might well be my last meal. And if we didn't crash, if I didn't die, I kept asking myself, would I be revisited by this food at fifteen hundred feet?

"In most planes, stability is a good thing," the other pilot was saying to David. Did that mean that gliders were not stable? I wondered. Flying in an airplane without an engine—this was definitely the most dangerous thing I'd ever done on purpose. And it *was* on purpose. I wanted to feel what it felt like to be carried along by wind. Or I had wanted to feel it. An hour, now, from takeoff, and I was no longer sure.

"Being stable requires less energy," the other pilot said loudly. He had an Australian accent and a confident manner. Gibo had told him I was going up for the first time, and I had the feeling that this speech about stability—a very calming word—was for my benefit exclusively.

"Monarchs are stable in their gliding configuration," David said. "If they have an active control system, their nervous system and muscles are going to be operating, and they'll use *more* energy."

This I understood. A monarch can carry only about 125 milligrams of lipids—its fuel—in its body. It takes just ten hours of powered flight—the kind of flying that is characterized by beating wings—to deplete that store.

"The idea is to get from here to there with as little energy used as possible," David said. "On the other hand, when they're attacked they have to go into violent maneuvers. They can flap and unflap their wings and beat them in different planes."

But what about us, flying in a craft with fixed wings and no fuel? What did we have to work with?

Not a lot, it turned out. At the gliderport David showed me our plane, a battered twenty-five-year-old Schweizer 2-33 trainer. Disproportionate to the airplane's body, like the arms of a rangy teenager that had grown faster than his torso, the wings spread out on either side of a remarkably small and compact hull. The hull was painted orange in a somewhat haphazard manner, the green and black of earlier paint jobs peeking out here and there. The steel housing was battered and pinged, and the effect was hardly reassuring. It looked like a jalopy.

The glider's wings were orange as well, and rounded on top—airfoil wings. This, I knew, would help keep us aloft. As air moved over the top of the wing, the airfoil would slow it down and disperse it. With more pressure below the wing than above it, the airplane would be pushed upward. This was lift. A monarch's wings were orange with black, too, but all similarities ended there. Butterfly wings were flat. Lift came from flapping, from churning the air until it created a whirling mass that moved along the leading edge of the wing.

The gangly orange sailplane, though not nearly as elegant as an orange-and-black monarch, had certain mechanical advantages over a butterfly. It had a rudder, located near the tail, that kept the fuselage aligned with the direction of flight. It had a pair of ailerons, one per wing, that controlled air-flow and offered lateral control. It had elevators on the tail to make the nose point up or down. It had spoilers to reduce lift. Each of these was available to the pilot should he wish to change the airplane's flight angle, its altitude, or its direction.

Professor Gibo was explaining this as we hoisted ourselves into the rudimentary and snug cockpit, me up in front, he in the back. Feet forward, I wriggled into place like a sausage being packed into casing. The plane—the inside of it, any-way—was narrow and tinny. There just wasn't much of it. And once the Plexiglas canopy was lowered, it seemed smaller still. Not a good place to be a claustrophobic, I was thinking, looking out the bubble overhead. Or to be a control freak, either, since there were almost no controls. Just the altimeter, rudder pedals, spoiler aileron, and elevator stick, and the tow-release knob to disengage the umbilicus connecting the glider to the tow plane.

"I don't need to know about any of these, right?" I called to David, looking for assurance that he, indeed, would be pi-loting the plane. But David couldn't hear me. The tow plane was buzzing up ahead and the yellow rope between us was los-ing its slack. After another second it grabbed the little orange glider as if it were a recalcitrant child and pulled it down the grassy runway. We clattered along, then lifted off the ground for a second like a kite on a short string, dipped back down, then took off again for real as the yellow rope stretched and grew taut. Twenty, fifty, one hundred feet and climbing. As I looked down at the receding ground, a line from a nameless poem went through my head: "From this there's no returning, none."

"Watch the tow rope," David called out to me over the loud, maddish complaint of an airplane being yanked through the air. "You'll know when we're going to hit a bump because you'll see it in the rope first." I guessed this made me feel more secure, though "more secure" might suggest that I felt *somewhat* secure, which at that moment I did not. We were bobbing around pretty regularly, mirroring the fits and starts of the tow plane but on time delay, like bad lip-synching. Still, knowing when it was going to happen let me tense up beforehand and brace myself.

"Five hundred feet," David shouted, a fact that I accepted ambivalently. Up was definitely better than down, but up meant we were moving farther away from the ground, from the world where gravity was as transparent and unthreatening as air. "Seven hundred feet!" I caught sight of the altimeter. It was inching up. But I didn't need an instrument to tell me that. The fields and farms below were growing distant, flattening the third dimension till it looked as if, really, there were only two.

As we approached a thousand feet, David asked me to call out the readings. "You have to release the rope at two thousand feet," he added.

"I do?" I yelled back. Apparently, I did. There was only one release knob, David explained, and it was by my knees. I put my hand on it and looked straight ahead at the rope, which was sending a wave in our direction. Bump.

"Fifteen hundred feet," I called. The higher we went, the more the sun bore through the canopy and spread heat and light relentlessly. No wonder David was in short sleeves and a wide-brimmed floppy hat. There was no escaping the sun. Off to our right I could see another glider, off its tow, making tight circles.

"It's in a thermal," David shouted, seeing it, too. "Don't worry. We won't hit that plane for the same reason you don't run into the car in front of you, or next to you, when you're driving."

Really? I knew plenty of people who had inadvertently rear-ended another vehicle. Didn't they count? "Seventeen hundred feet." The climb was steady, and the horizon stretched in front of us, and with no landmarks there was no way to distinguish eighteen hundred feet from nineteen hundred feet. But we had reached nineteen hundred feet and were pulling up to the invisible station. My hand tightened on the release knob, independent, it seemed, of my reluctance to separate from the motor that was carrying us aloft. Something about letting go felt suicidal, like pulling the switch on one's own electric chair.

"Two thousand feet," I sang out to David.

"Two thousand feet," he confirmed. I pulled back on the knob, which resisted for a moment, then gave way with a jerk and a loud pop that startled me. The yellow cord snapped away, and all the noise did, too. Quiet rushed in around us, a whispering quiet, and the plane untensed like a hand gone from fist to open palm, and for the first time since taking to the air we were flying, really flying, and it was, much to my surprise, glorious: serene, buoyant, unlikely, glorious. The air, which until then I think I had not seriously taken into account as a force in my life, was holding us up. I could feel it, as if it had especially large fingers that pressed into my flesh. Air was a fluid, David had told me earlier, and now, buoyed by it in a tangible way, I felt what he meant. He pushed gently on the right rudder pedal and we turned slightly, then rushed forward in a graceful swoop.

"The clouds are pretty fragmented," David said, referring to the thermal updrafts we were seeking. David's voice was calm and measured. He was a seasoned pilot. I figured if he was untroubled by the fragmentation, I should be, too. I leaned back in my seat and looked out the window. I could no longer say that I didn't like to fly.

A MONARCH TRAVELING to Mexico from where we were in southern Ontario would have to fly about two thousand miles to reach one of the overwintering sites, and it would take it ten or eight or six or fewer weeks to get there. Powered flight—flapping—would propel the bug forward, but at a cost in terms of both fuel consumption and time, since powered flight, despite its name, is relatively slow. When most people think of monarch butterflies' migrating to Mexico, though, and then going back to the United States and Canada, that's the image they have in their heads—a small and fragile and vulnerable creature flapping like mad in a determined and enervating manner.

Up in the cockpit of his Grob 103, David Gibo understood in an intimate and practical way that a butterfly could not possibly flap its way to Mexico. Gibo could travel on the wind for hours in his fixed-wing airplane, soaring and gliding across the sky using thermal updrafts, or rising columns of warm air. The butterflies, he reasoned, must do the same thing. Gliding (nonpowered flight in still air) and soaring (nonpowered flight in moving air, such as a thermal) were the methods that enabled both him, in his Grob, and a monarch, with its aerodynamic constraints, to travel long distances with little wear and tear and even less fuel consumption.

"Typical gliding and soaring flight, with the wings outstretched, [does] not generally require any effort from the insect," noted the German zoologist Werner Nachtigall in his book *Insects in Flight*. "No energy is expended, either because the wings themselves automatically take up the gliding position as soon as the flight muscles are relaxed . . . or else [because] there is a 'click mechanism' which puts the wings into the right altitude mechanically. This latter method is how butterflies assume a gliding mechanism."

It was in the same book that, shortly before heading up to Toronto to meet David Gibo, I had read this: "[A glider pilot] climbs steeply in one thermal, then leaves this and glides obliquely downwards towards the nearest thermal, in which he soars again, and so on. By this means he can make long, cross-country flights without having any engine. The only snag is that each time he must find another thermal before his downward glide has taken him to ground level. If he does not he must land."

Up above the farms of Wellington County we drifted with lazy purpose, circling laconically and moving in a direction that felt like up. I knew that some gliders could course the sky all day, and that some had reached altitudes that required supplemental oxygen. Our goal was to stay aloft as long as possible, which meant at best twenty or twenty-five minutes. I also knew that sometimes the thermals were so prevalent that it was hard to come down, though this was rare, and that most of the time pilots crossed the sky as if the thermals were stones in a stream—upwardly mobile stones that moved them both up and over at the same time. Lose the thermals, and you were on your way down.

"We're in sink!" David said from the backseat. His voice was steady, uninflected. My eyes fixed on the horizon, trying to see what he meant.

"What is 'sink'?" I asked, seeing nothing but wispy clouds and another glider, far enough away to look smaller than it was. David adjusted the rudder and the aileron. The nose of the plane, which was pointing downward, wiggled up.

"It means that we've lost the thermal and we're losing altitude," he said, but sounding calm and businesslike. *Sink.* It was a delightfully descriptive word. I knew I should feel scared, but I didn't. I checked the altimeter: we had dropped five hundred feet within a matter of seconds. Even so, we were already on our way to another thermal, banking sharply as we aimed for a destination in the sky that only David Gibo could make out. In less than a minute we had caught the updraft as if it were the 1:17 commuter train pulling out of the station.

"Usually what you do is fly a pattern in order to fly into a thermal," Gibo was saying. "Now, with the butterfly, it goes in the direction it wants to go, finds the thermal, and then behaves appropriately."

I can't say I was listening attentively when Professor Gibo explained this, or when he mentioned that meteorologists had picked up migrating monarchs on radar, flying at five thousand feet. Instead I was thinking about the words *We are in sink,* and the fact that we had been falling from the sky and that the sky—invisible and elusive—had held us up.

•   •   •

ON CLEAR DAYS, days without rain or significant cloud cover, monarchs start dropping out of the sky at around three in the afternoon. There's plenty of sunlight left then for flying, but sun, for navigational purposes, and warmth, for thermoregulation, are only two of the factors that keep a monarch in the air. Another is wind—which way it's blowing, if it's blowing, and whether it's moving horizontally or vertically in a reliable, exploitable way. In the late afternoon, as the earth cools and the thermals break up, monarchs lose their free ride. They drift down to the ground, flapping as they need to and adjusting their glide angles, sinking yet held aloft by their thin, rigid wings.

Our airplane, despite its superior aerodynamic design, had the disadvantage of weight. It, and we in it, were far heavier than the wind. Gravity was always a factor. We caught a small thermal, gained a few hundred feet, lost it again, dropped a few hundred feet, found another updraft, gained fewer feet than we had lost, found another thermal. From the moment we came off the tow and for the whole time we were flying, we were in the process of landing, moving closer and closer to the ground until David said we were getting to the point where we'd have to put down. From the air we could see the gliderport with its stable of aircraft and its tin-roofed hangar reflecting the midafternoon sun. David said we had to fly a left circuit, over and behind it, then approach from the southeast. The trick would be in catapulting the power lines, since the wind, as we got closer to the ground, was gusting unpredictably in a southwesterly direction. We could see it now, combing the trees, which were swaying. "Not ideal circumstances for your first glider landing," David said flatly, plying the rudder.

Landing scared me in a way that flying hadn't. I had gotten used to the notion that air was a fluid. I understood, in a visceral way, the metaphor of swimming that David liked to use. "A monarch swims through the air," he said, an image that I both understood and found comforting. But landing required a different metaphor, and since none seemed to be on offer, the hard and solid reality of the ground seemed even more solid and harder. We passed over the hangar, then banked near the road, turning toward the field. David pressed on the rudder again and the nose of the airplane made a steep angle to the ground, one that my internal protractor found terrifying.

"Here we go," David said, and the glider began to accelerate, heading directly for the power lines. Too frightened to close my eyes, I saw the cables slip under the belly of the plane with what looked to be about two feet to spare as the wind buffeted us this way and that. I had seen monarchs tossed about like that, fighting to fly southwest as the wind pushed them to the east, till their bodies were moving eastward but their heads were aiming in the direction of Mexico, as if will alone would get them there. David held hard to the controls, wrestling the wind with his intelligence. It needed only to be a draw, and it was, as we came in over the runway and leveled off. Twenty feet, ten feet, one foot above the rutted tarmac, dropping with a thud as a group of air cadets ran out to catch us, holding on to the wings of the plane to pin it down before the wind could do it damage. We were down. David congratulated me. I thanked him. We popped open the canopy and climbed out and for the first time all day caught sight of a monarch butterfly flying at knee height, flapping energetically. Although it was facing southwest, it was moving southeast. Professor Gibo and I paused to watch it struggle against the wind for a time, only to give up and park itself in the grass. "It was losing ground with respect to the overwintering sites," he explained as we went into the airport office to sign up for another flight. "Natural selection has produced an insect that can detect that. So it went down to where it could control its flight. If you watch, you'll see that happen a lot."

AS A BIOLOGIST, David Gibo watched a lot. With his binoculars and his sailplane, he was an entomological voyeur. The problem was that watching, as an individual, wasn't getting him the information he wanted. What he was getting on his own was too sketchy, too thin. Gibo was interested in the monarchs' flight tactics—in how, specifically, they exploited the wind in order to find their way to Mexico and back. And he was interested in the vectors they used and the directions they flew in. He knew that though their general orientation as they moved overhead in Toronto was southwest, they did not fly in a straight line. Not that they wouldn't if they could, but the wind did not offer such direct routes. And he knew that for them, flying was like sailing: it required tacks to be made to close the distance between where they were and where they wanted to be.

"You can only do certain things when moving through the air that will allow you to get to where you want to go," he said the day after our excursion to Arthur, as we were walking from his office on the University of Toronto's Erindale Campus out to the parking lot, which serves as his study site. "This is the limitation of physics, basically, because, as you know, you are in a fluid. And you are using your interaction with this fluid to keep aloft. And if you're going to take advantage of things like thermals to get a free ride and increase your potential energy so you have something to play with, then that means you're going to have to have pretty darn complex tactics, because the way to get up is to circle this rising air, which is drifting downwind and probably not going in the direction you want it to go. And now you have to compensate for the displacement due to circling in a thermal, gaining altitude. The fact that wind increases as you go up, the fact that you swing clockwise as you go up—you've got all those things going on, and you have to decide which direction you're going to fly in when you leave that thermal. That is a fascinating problem—what the monarchs are doing in northeast winds, in east winds, in southeast winds—and something I'd like other people to get dragged into."

The Erindale Campus, where Professor Gibo taught was a small branch of the university in Mississauga, a suburb about half an hour outside the city. It was pleasant if undistinguished, with the look of a large suburban high school. Because many of the students commuted, parking lots dominated the grounds, circling the main academic building with a broad band of blacktop. Gibo walked over to his car and laid out his instruments on the trunk, using it as his lab table—a habit, I'd come to notice, common among field biologists. "Any fool can do science," he said, taking stock of his kit: field glasses, wind gauge, thermometer, compass, logbook. "That's why it's so powerful."

Gibo was a proponent of laypeople doing the looking, the measuring, and the recording essential to empirical science. "Every single comet was discovered by an amateur," he said. "You want to know why? Because they're the only ones who have the time to stand out there, take pictures, and then go back a week later to see if anything has moved. People are smart. They can do this work."

The work to which Gibo was referring was the campaign he had launched on the Internet to enlist people to collect data about the flight patterns of migratory butterflies. "However clever their flight tactics, however mysterious their method of navigation, everything has to resolve into a series of simple rules," he told the volunteers. "Lots of rules, hierarchically arranged and nested sets of rules, but above all, simple rules. The tiny nervous systems of butterflies just aren't capable of anything else." The rules would be revealed, Gibo believed, if enough data could be collected and then analyzed.

To encourage this, Gibo had set up an extensive Web site that was part field-biology cram course and part public record book. Participants were asked to record, in a standardized format, the flight behavior of migrating butterflies, including their estimated altitude, the type of flight (flapping, gliding, soaring), the wind velocity, the ambient temperature, cloud type, and heading. They were also asked to note—that is, to interpret—the context of the observation. "Butterfly was apparently engaged in courting behavior with another *D. plexippus,*" Gibo himself observed from the parking lot on September 1, a day when half the sky was dotted with cumulus clouds and the ambient temperature rose in the half hour he was out there from 20 degrees celsius to 23.5. "The one being observed appeared to maneuver to keep in tandem to two hundred meters, then glided back down to disappear behind the South Building. Two more solitary *D. plexippus* were also seen. Butterfly started at three hundred meters and soared level before gliding down to two hundred meters. Its descent was gradual. A gull was also seen in field of view. Butterfly flapped upward. A second, lower *D. plexippus* flew past. Two more *D. plexippus* were seen. One glided down to a group of trees, one was flapping SW at five meters."

Two weeks later a glider pilot in Worcester, Massachusetts, added his observations to the record, noting that during a routine flight at midday on September 16 he had encountered two monarchs at about forty-five hundred feet, soaring in a thermal that was also host to some fifty hawks. Again, cumulus clouds were scattered across the sky—often a good indication of thermal activity—and the wind was moderate. The thing that most intrigued David Gibo about this report was a detail that he posted as an addendum: the butterflies had been about sixty miles inland, and with the wind drifting in a southeasterly direction, they'd been moving with some dispatch toward the Rhode Island Sound and the Atlantic Ocean. By simply circling in thermals, Gibo surmised, they'd reach the water in just over four hours. If they spent half their time circling and half moving straight ahead by flapping and soaring, they'd get there about an hour earlier. Once they got there, Gibo said, "they would probably continue roughly southwest, paralleling the coast, pass through Cape May, New Jersey, and [maybe] even be counted by [Lincoln Brower's associate] Dick Walton at the Cape May Bird Observatory."

This, of course, was speculation. There was no way for anyone to know where these particular butterflies were going to go, or where they were going to end up. Gibo was making an educated guess based on what he knew about wind speed and cloud formation and monarch butterfly navigation. To me it was a story, but a compelling one based on some real things, the way certain literary nonfiction both is and is not true. To David Gibo, though, the story was the essence of a certain kind of scientific enterprise, wherein the weight of reality—the mass of observed phenomena—allowed the creation of hypotheses (stories themselves) whose usefulness, if not their truthfulness, could be tested against the further accumulation of data.

"Most of science takes place in people's heads, and then you present an argument to convince people and of course you arrange your argument in a way to make it maximally convincing," he said. To me it sounded remarkably like plotting a novel.

OUTSIDE IN the Erindale parking lot, the sky was conspicuously free of bug life, at least at an observable level—a result, perhaps, of the wind, which was gusting. We were looking straight up, heads bent back as far as they could go, and nothing, nothing, nothing was in our line of sight, and then a kestrel. "This bird is gliding, by the way, it's not soaring, and it's going between thermals," David said. "It just picked up a thermal," he reported a minute later. "See it circling? See the circle? It's going in a circle and gaining altitude."

"When we were up there circling, we were playing," I said. "When nonmigratory birds ascend in thermals, are they playing?"

"I doubt it," David said, keeping his eyes fixed overhead. "They're looking for food, checking out their environment, being shoppers, watching for other individuals, displaying to their mates, keeping cool. Crows, yeah. Crows tease."

On this day, though, the butterflies were teasing, having promised, by showing up in large numbers in years past in this part of the world in this part of August, that they would be here now.

"Despite all the observations you've made here over the years, empiricism seems no more precise than divination," I said to David when still, half an hour later, not a single monarch had flown by. I was hot, my neck hurt, and my eyes were going blurry. I was ready to pack it in.

Gibo did not for a second loosen his posture. "We need more data," he said. As if that would be enough. In the meantime, for him, the middle of the story was just as compelling as its end.

Chapter 9

AT JUST ABOUT the same time that the monarch spotted at forty-five hundred feet over Worcester, Massachusetts—the one David Gibo surmised was on its way to the Atlantic coast at Rhode Island—would have been in sight of Cape May, New Jersey, I was getting there myself. Cape May Point, a spit of land that looked from certain perspectives as if it were rudely elbowing the ocean, had long been a draw for birders, who gathered there in great numbers, especially in the fall, when the sheer numbers of migrating songbirds, raptors, and water fowl could darken the sky—or at least seem to. Nearly four hundred bird species had been identified there. The checklist I was carrying in the back pocket of my jeans listed 391 in all and noted that John James Audubon had spent time in the nearby swamps and marshes, executing some of his well-known bird studies. The

Cape May Migratory Bird Refuge was at the joint of the elbow, where the Atlantic Ocean and the Delaware Bay mingled. Farther inland, but not far from the water in any case, was the Cape May Bird Observatory, one of the most active ornithological education centers in the country. Walk along the trail to Higbee Beach, stop at Stone Harbor or the Seventh Street Sea Watch, and others would be stopped there too, eyes trained upward or fixed on a particular tree limb. Stroll down Beach Drive, the wide esplanade fronting the ocean, and among the sunbathers and kite fliers and fudge eaters, would be an uncommissioned army of birders, easily identifiable by the Leica Trinovids and Bausch & Lomb Elites hanging on neoprene straps around their necks—benign but expensive ordnance.

Lepidoptery is a lesser franchise on Cape May, but a growing one. There are 105 known butterfly species on the cape, and its peninsular shape has made it a refuge for migrating monarchs. Birders looking for passing hawks often picked them up in their spotting scopes, in numbers that made them seem like winged rain.

Dick Walton was one of these. The author of a number of books that were part of the ornithological canon, as well as a series of popular birding-by-ear instructional tapes, he had gone down to Cape May from his home in Concord, Massachusetts, on October 10, 1982, to witness the legendary hawk migration. It did not disappoint. On that day 2,622 sharp-shinned hawks, 62 Cooper's hawks, 50 peregrines, and 130 merlins passed overhead. But Walton was seeing something else as well, as he observed in his journal: "All day long we witnessed a phenomenal monarch migration. The butterflies were as constant and continuous as the hawks." And it wasn't just that year. The pattern—hawks and monarchs migrating on the same wind—continued in subsequent years, piquing Walton's interest. Dick Walton is a free-lance natural historian and, though primarily a bird-watcher, he was not content simply to look at the butterflies going by.

"In the fall of 1990 I decided to spend two weeks in Cape May with the idea of planning a long-term research project on monarch migration (what the heck—somebody has to do the work)," he recalled. "I tried out several census methodologies at various places on Cape May Point. At Sunset Beach on September 27, in eight one-minute observation periods, I counted 618 monarchs, for an average of 77 monarchs per minute. The following day I counted over a thousand monarchs streaming through Cape May Point State Park. Although there were plenty of monarchs, there were also many puzzles. One of my first discoveries, while counting monarchs at Cape May State Park, was that the direction of the migratory flight seemed to reverse itself for no apparent reason. At one point a steady stream of monarchs would be heading southwest and then within the space of ten minutes, the whole flight would be going northeast. When I returned home I puzzled over the data, and even though there were more questions than answers, I was convinced Cape May would be an ideal site for my study. So I wrote Lincoln Brower that December and we began discussions about setting up the project."

The project, though billed as the Monarch Migration Association of North America, was really just Dick Walton, an assistant, a car, and a hand-held, nonautomated counter. The idea was to establish a census route through Cape May that would be followed three times a day in order to begin amassing data on the numbers of monarchs migrating through Cape May each fall. While the figures for any one year would be interesting in and of themselves, it was the accumulation of data that would be most telling. "Good" years and "bad" years, "big" days and "small" ones—all would settle out, or become revealing, over time.

It was this data, eight years out, that brought me to Cape May on one particular day. Historically, September 19 was a big day there, a day on which in previous years thousands of monarchs had flooded the skies and clustered on the trees and nectared in the pocket parks. If the data were any guide, September 19 was the day to be in Cape May—especially, I reasoned, if the butterflies seemed to be scarce elsewhere. Either they'd be here, and that would be significant, or they wouldn't, and that would be significant, too. It was win-win, though I'd rather prevail by witnessing a deluge of monarch butterflies.

DICK WALTON MET ME at the hawk watch bench at Cape May Point. The bench is a raised platform bordered by railings, equidistant from a marsh (good for sighting water-fowl), the beach (good for sighting shorebirds), and the parking lot (good for sighting the latest in ornithological gear). Most days it fills early, as curious tourists and serious birders sit shoulder to shoulder looking into the sky. Even among these, Dick Walton was recognizable. Bearded, with a worn Red Sox cap on his head, sporting an ancient pair of binoculars, he looked like he belonged there.

"We do three census runs per day, at nine, twelve, and three o'clock," he told me. "They take fifteen minutes each.

We note the date, the temperature, the cloud cover, the wind direction, and the wind speed, determined by leaf movement. It's pretty standard, and it's fairly subjective. Any observation is.

"On the census runs we're not trying to count all the monarch butterflies in Cape May. We're taking a sample. If you sample along a route long enough, it will become more accurate. You might see thousands over at the hawk bench, and up on the dunes, but this is more accurate. There you're getting flight reversals. You're counting the same butterflies more than once. They're deciding whether or not to cross Delaware Bay—it's eleven miles. We know they cross open water because we see them when we take the ferry. But the wind has to be right or they turn back, which is when you start to miscount. Our census spreads it out so we get a better idea of what's really going on. If you want to census dogs in New York City," he concluded, "you wouldn't want to go to a fire hydrant."

It was not quite nine in the morning. We drove out of the parking lot toward the Cape May Lighthouse, then turned up Lighthouse Avenue and headed for the Higbee Beach Wildlife Management Area on the bay side, where the census began. Dick pulled up to the parking lot, which was already full—it was a hot spot for birders—and then he turned around and consulted his watch. The census was due to begin on the hour. And it would have, if twenty people hadn't been stationed in the road, looking for a particular warbler in a particular tree. We eased out slowly, picking our way around them and then working up to a steady twenty-five miles an hour, the uniform census speed. Some days Walton was followed by another person in another car making her own, independent count, but not today. It was just the two of us, and the rule was that we were not to tell each other when we saw a monarch. We were not to twitch or make any sort of movement that might give it away. Dick Walton held his counter out the window on the left side. I held mine out the right. He didn't want either of us to influence the other.

"Most butterflies disappear when it's cloudy," Dick said. He was driving one-handed, scanning the sides of the road, the middle distance, the sky, his eyes moving quicker than his car. "Once I was doing the census in a nor'easter and saw one blowing down the beach. It didn't want to be flying. I counted it." He turned onto Bayshore Road, hugging the shoulder, letting the other cars pass.

"Sometimes I think I should have a banner so people know why I'm driving like this," he said.

I was concentrating on the farmland we were passing, looking hard for monarchs and seeing none. I thought I had noticed Dick tapping his clicker, and I wondered if I had missed one, or more than one. It was making me anxious, tetchy.

"Did you know that Cape May is the lima bean capital of the world?" Dick asked as we passed a processing plant called the Beanery. No, I didn't. "Seaside goldenrod is a good source of nectar," he said a minute later. I wondered if this was a clue and scanned the plants as we passed by, but still I saw no monarchs.

We turned off Bayshore and drove in the direction of the bay. So far the big day wasn't happening. Even my poker-faced driver seemed disappointed. "Negative data are good," he said, "but they're not much fun to collect." Maybe he hadn't seen any butterflies after all. We drove past the lighthouse again. "We just need a little more brightness," he said, "but I don't think we'll get it on this run." Click. I heard his counter tally a butterfly, just like punctuation. I looked around madly, didn't see a monarch, and left my counter registering a row of zeros. Now we were zigging and zagging through the residential neighborhood between Lily Lake and the sand dunes. It was seventy-three degrees and overcast. A monarch was crossing the road. "There, look!" I said, waving my hands excitedly. Dick Walton shot me a bemused look. So much for "double blind." I had given it away. I heard his counter click again.

It was 9:17 on the morning of the big day. Total number of monarchs sighted: two. We returned to the hawk watch bench, where Dick Walton gave me an impromptu lesson in identifying birds in the sky. There were thousands of tree swallows sailing about, with a big bird laboring among them.

"Tell me about that big bird," Dick Walton said. "What do you notice about it?"

"It's big," I said.

"It's an immature bald eagle," he said.

"Where?" said a chorus of voices. The hawk watchers turned from what they had been doing and searched for the eagle.

"This is great," said a portly man in shorts and a photographer's vest, sporting a Cape May Observatory hat on his head. "I didn't look at birds till I retired here," he said to me. "I used to think people who looked at birds were nerds. An immature bald eagle—haven't seen too many of those. Wish I had started doing this a long time ago."

Dick pointed above a stand of cedars. "Now there's a good bird," he said. Again, as if on point, the birders turned and tried to pick up what he was seeing. "Can you tell from the shape of the wing what it is?" he asked me. I couldn't.

"It's an osprey," he said. Another bird of prey. And there, at a hundred feet, was some prey, a lone monarch gliding in exactly the same direction the wind was blowing.

IN ADDITION TO creating a historical record of monarch butterfly behavior at Cape May, Walton's Monarch Migration Association of North America (MMANA) sought to do something else as well: to demonstrate, once and for all, that the movement of monarch butterflies down the East Coast of the United States was not an aberration; to show, through numbers and constancy, that there was a real and established Atlantic flyway that led to the Mexican overwintering sites. This had been a matter of dispute for at least a decade, beginning with Lincoln Brower, then a professor at Florida, squaring off against—again—Fred and Norah Urquhart. It was the Urquharts' understanding, based on their tagging data, that monarchs found on the East Coast had drifted there unwittingly, the victims of winds that carried them southeast instead of southwest.

After the first four years of the Cape May census project, it looked as if the evidence would eventually prove the Urquharts wrong. As Brower and Walton wrote in the *Journal of the Lepidopterists' Society* after correlating their findings with data from the North American Butterfly Association's annual Fourth of July butterfly count, "During the four-year period of our study we have consistently recorded large numbers of monarchs at Cape May, New Jersey, [and] have also regularly observed migratory behaviors including 1) mass movements

along beach and dune lines; 2) a high degree of directional-
ity . . . ; 3) roost formations; and 4) significant build-ups and
exoduses on consecutive days. Another notable characteristic
has been the timing of the fall movement. In each of the
four years studied, the numbers of migrants peaked during
the third week of September." Therefore, they concluded,
"our Cape May observations argue in favor of describing the
Atlantic coast migrants as routine constituents in the monarch's fall migration. The numbers and behavior of monarchs
observed leave little doubt that a significant migration has occurred at Cape May in each year of our censuses. Aspects of
the timing of the migration, in particular the recurring September peaks, also indicate a routine passage of monarchs.
Such consistent timing would be unlikely if they were caused
solely by weather conditions such as cold fronts, because the
latter do not occur at the same time each year. Finally, the
correspondence of the Cape May and 4JBC [Fourth of July
butterfly count] data sets suggests that the number of monarchs passing through Cape May is representative of northeastern breeding populations as a whole. If this correlation
holds in future years, it will strengthen the hypothesis that the
Cape May migration is representative of the population of
northeastern monarchs, rather than comprising an 'aberrant'
group displaced by atypical weather conditions as hypothe-
sized by the Urquharts."

Now, with four more years of research completed, Brower
and Walton's conclusions still held. While the numbers of
monarchs passing through Cape May fluctuated from year to
year, their presence was constant. Monarchs could be seen up
in thermals with the migrating hawks and roosting in the
trees around town. If in decades past their aggregations had

been merely the subject of anecdote—the late Roger Tory
Peterson, for one, "recalled from visits to Cape May in the
early 1930s trees so completely covered with monarchs that
they were 'more orange than green' "—the MMANA census data lent those observations weight. ("Those folks lucky
enough to have been in the Cape May area on either September 19th or 26th were treated to a blizzard of monarchs,"
Dick Walton reported in his annual summary for 1997. "In
fact, on the 26th, 652 monarchs were counted on the three
census runs. This eclipsed all previous records for daily run
totals. Observers at the [hawk bench], Higbee, Cape May
Meadows and even along the streets of downtown Cape May
reported hundreds, even thousands, of monarchs on both of
these days.")

But that was a year earlier. On this September 19, a casual
observer would have been hard pressed to discern a migration at all. All day, it seemed, we were busy looking through
our binoculars from the hawk bench, or inching our way
down New England Avenue, collecting negative data.

And that day, it turned out, was not unusual. Throughout
the 1998 migration season, Dick Walton and his associates
noted that the number of monarchs passing through Cape
May was less than half what it had been the previous year.
Walton attributed the paucity to the Mexican drought and
early storms in the northern breeding range. Still, he couldn't
say for sure.

What he could say, though—finally and, at the end of
1998, conclusively—was that the Urquharts had been wrong.
For the first time since 1992, when MMANA began tagging
monarchs, butterflies tagged at Cape May had been recovered in Mexico, at the El Rosario colony. And not just one or

two—*seven* had been found there. But it got even better than that: four Cape May butterflies had also been recaptured farther down the coast, in Virginia. The data were like those chemicals that reverse invisible ink: once they were applied, the Atlantic flyway that Lincoln Brower and Dick Walton had always assumed was there at last revealed itself.

AT SOME POINT, of course, the monarchs that flew down the eastern seaboard and then to Mexico had to swing inland. Maybe they turned in the Carolinas, or maybe they hugged the coast to the thumbnail of Florida, where, confronted with open water and its maniacal choice, ocean or gulf, they chose the gulf. Whatever their course, there was still no evidence to support either notion, just the tantalizing fact that some coastal monarchs found their way, somehow, to El Rosario.

"It's a real enigma, the flow of monarch butterflies down the East Coast," Chip Taylor said to me again one day when I was visiting the Monarch Watch office in Lawrence. "Why do we have these big holes?" He pointed to the map of tagged monarch recoveries, the one that had no red pinpoints anywhere in the Southeast. "The longitudinal data are great, but the recoveries don't meet up with our expectations regarding latitude." He tapped the hole in the map with his pen. "The butterflies ought to be turning right."

It was the Big Right Turn theory again, a conclusion the data merely hinted at. David Gibo had gone in search of its coordinates in his ultralight airplane and failed to find them. Chip Taylor had distributed hundreds of thousands of Monarch Watch tags, but none had ever shown up in Mexico from any southeastern state. Now he was on to something else. Or rather, his colleague Sandra Perez was. The year before she had taken Kansas butterflies to Washington, D.C., and Dalton, Georgia, and released them to see which way they went; all had behaved as if they were in Kansas, flying southwesterly. Now she was going back to repeat the experiment, but with a difference: this time she was going to cage the monarchs in situ for a few days, on the theory that they might need a little time in their new neighborhoods to figure out how to read the local cues.

"I still expect them to behave the way they would if they were in Kansas and go southwest, but Chip expects them to adjust for their new orientation and fly more westerly," Perez said. More westerly would suggest the Big Right Turn.

As she talked, Perez took a quick inventory of the materials arrayed on the table in the biology annex, located on West Campus, in a trailer she shared with a postdoc who studied crickets: plastic hangers, toothbrush holders, duct tape, pieces of wood. Perez was making artificial nectar feeders for the butterflies she planned to keep in captivity. For as long as they were with her, they'd be treated to a cocktail of Gatorade and protein, served from a soaked, colored kitchen sponge. The feeders held the sponges.

"What I love about field biology is that I can get all my equipment at Wal-Mart," Perez said, bridging two hangers with a piece of wood that she then secured with a length of duct tape. That done, she taped the top of a toothbrush holder onto the wood. This would cradle the sponge.

"When the migrants come through I collect several hundred at a time," she said, tearing strips of tape and hanging them off the side of the table. "This many feeders is probably

excessive, but I don't want anyone to be hungry in there."
She motioned for me to start taping the wood to a hanger.
Production swung into high gear.

"I'm sure one of the biological supply houses sells something like this premade," she said after a while, "but this is a
lot cheaper. And a lot more fun."

BUT IF THE HANGERS, the toothbrush holders, the
sponges, and even the tents where the captives would be
housed could be bought at Wal-Mart, Perez was still missing
the one thing she could not buy there: butterflies. Specifically, nonreproductive migrating butterflies. Migrants were as
scarce in Lawrence that fall as they were in the Adirondacks
and Toronto and Cape May. The Baker Wetlands, not far from
the university campus, a reliable roost for migrating monarchs in years past, was now as empty as a condemned building. We saw just twenty of them in a quarter-mile stretch on
which on an average day the year before we would have seen
ten thousand. That order of magnitude could not be explained away. "They've never been this late before," Chip
Taylor said of the monarchs. But he couldn't say where exactly they were, or where they'd be coming from.

He had a hunch, though. When you needed migrating
monarchs in Kansas, there was one sure place to go: Wamego.
There was a high school biology teacher there, Terry Callender, whose students had tagged more monarch butterflies than
any other group. One year, twenty or so kids had captured,
measured, weighed, tagged, and released an unprecedented
twelve thousand monarchs. They'd had a lot of recoveries, too.
Terry's classroom, Chip said, represented the best of what

Monarch Watch had to offer students. If they had monarchs there, he was sure they would share. Sandra and I consulted a map. Wamego looked to be about two hours away, between Topeka and Manhattan, on Highway 24. We gathered nets, extension poles, envelopes, and high-beam flashlights. The high beams were key, Sandra explained, since we'd be jacklighting the butterflies.

"Isn't that poaching?" I wondered aloud.

Sandra looked at me with amusement. "This is *science*," she said, laughing.

TERRY CALLENDER, an affable, bearish man with a neatly trimmed beard and gray hair, waved us into his classroom in the basement of Wamego High. It was the end of the day, and students were gathered by his desk waiting to sign out butterfly nets while Callender, wearing a faded Monarch Watch T-shirt, teased them.

"If you want to get out of science process class on Friday, bring in a thousand monarchs and we'll have too much to do to do anything else," he told them. They nodded earnestly.

Terry turned to us. "We usually do five or six hundred per class. They get really efficient, especially if we shut down the class. Sometimes a single student will bring in five hundred or even a thousand butterflies. For some reason we always get a lot of monarchs here. It can get really wild. Two years ago we had eight hundred we had to feed because it was too cold for them to fly. Sometimes we get so many butterflies that I don't teach the entire day and we just tag the whole time. It's like a factory. When school started this fall the kids came in and said, 'When do we get to tag?' I'd never heard that before. They are really into it."

On the wall over the sink in Terry's classroom was this message: "I hear and I forget. I see and I remember. I do and I understand." One look around and it was obvious that nothing
could better describe his educational philosophy. In a run-of-
the-mill school in a rural midwestern backwater, Callender
was pioneering the kind of hands-on, experiential learning fa-
vored by progressive educational reformers in more sophisticated places. There were waders piled up in one corner so kids
could explore the creeks that ran to the Kansas River, which
flowed alongside the town, and piles of butterfly nets, and rows
of microscopes, and a freshwater aquarium. A paper wasps' nest
hung from the ceiling, and there were animal skulls and duck
decoys. Terry Callender was a hunter, and it showed—not just
in the veritable museum of taxidermy that also crowded the
walls of his classroom, but in the patient, observant, expectant
attitude he brought to the study of wild things.

"Divide yourselves into groups of threes and fours," he
told the students when they had returned to their seats. "I'll
be right back." And he was, carrying a box of envelopes he
had just taken out of the refrigerator.

"OK. You are going to put a tag on each butterfly and
write down its number in the log. Then you are going to
write down whether it is male or female. How do you tell
the difference?"

"Something about the veins," one boy offered.

"Well, yes, there is something to that," the teacher said,
"since the veins tend to be thicker on the females than on
the males. But there's something else. Look at my shirt." The
kids stared at his shirt, which meant staring at his substantial
middle, which made a couple of them giggle.

"That one," a girl with long blond hair and braces said,
pointing to one of the butterflies. "It's a male."

Terry Callender beamed. "Right," he said. "See the dot? It's the pheromone sac. There's some controversy about whether it's functional or not, but only the males have them. Right. OK. When you're doing the log, I need you to write down wind speed—it's calm today; temperature—it's in the mid eighties; and the wing condition on a scale of one to four, where one is fresh and four is almost transparent because the scales have fallen off. When you're done, call for a runner—I need volunteers—and give your monarch to them and they'll go outside and release the butterfly."

The students broke into groups, girls with girls, boys with boys, and before long the room was percussive. *It's a one. Male. I think it's a two. Male. Female. Two. No, one. Runner, we need a runner. Female. Female. Two. Runner.* The bell rang. A few kids with buses to catch took off. A bunch of others stayed. It was last period. The halls were filled with students eager to leave the building. Callender's students remained in their seats, sexing the butterflies. "They just can't seem to get enough of this," he said.

That Terry Callender's students even had any monarchs to tag was evidence, he said, of their interest in the project. In past years the butterflies had been plentiful in Wamego— he recalled sending his own six-year-old into the front yard with a net, and the boy's returning a few minutes later with over a hundred monarchs—but this year they were scarce, coming in sporadically, not hanging around.

"This is the first time in four years that they haven't been here on the tenth of September in big, big numbers," he said. "Four years isn't a long time, but still, I think we were seeing a trend." We made plans to meet up again later, after the sun went down, in the city park.

• • •

WHERE I CAME FROM, jacklighting was the lazy way to bag a deer. Guys would troll the back roads in their pickups, shining bright lights into the edge of the forest, hoping to catch the reflection of a buck's eyes. The light would pin him there, and the guys wouldn't even have to get out of their truck. They'd just rest the rifle on the window ledge and shoot.

Jacklighting monarchs, Sandra Perez assured me, was nothing like that. For one thing it wasn't illegal. For another, it wasn't the light that paralyzed them; it was the temperature. They were already unable to move before the light was shined on them. This, it seemed to me, was a debater's point.

"We're going to shine the light under the branches and look for roosts," she said, ignoring my objections. "Sometimes they're hard to see. You have to be careful not to miss them because they blend in so well."

"It doesn't look good," Terry Callender said when we met up with him in the park. It was nine o'clock, and the reason he thought it did not look good was that his students had been out on the streets of Wamego for an hour and a half; he was pretty sure the town was already picked over. But then we walked a little farther into the park and Sandra shined a light on the underside of a silver maple and there, as if she were projecting it, was a small cluster of monarchs, crowding the branch like rush-hour straphangers on the Broadway Local.

"We'll snag them later," Terry said, hustling us out of the park and down a pleasant residential street of small houses with big porches. "We need to get to Mrs. Bradford's yard before the kids do. There are always monarchs there. Why that yard? Why those trees? There is no apparent reason. We tried to determine if they prefer one kind of tree to another, but they don't seem to."

We walked briskly, the sound of our footfalls seeming to precede us. There was no traffic; the streets were quiet, the houses welcoming. Muslin curtains, yellow light, roofs over-hung by old, old trees.

Mrs. Bradford's trees looked like all the others. We in-spected the one out front, then walked around to the back of the house. Terry Callender waved to Mrs. Bradford, who was on the phone inside. She hung up and came out just as San-dra was demonstrating her net-swiping technique. It was very smooth. Thirty monarchs tumbled into her clutches.

"Nice," Terry Callender said.

Some of his students appeared just then, walking out of the shadows as if they'd been summoned by an inaudible buzzer.

"What did you get?" they wanted to know—meaning "how many?"

"They're keeping score," Sandra whispered.

"Where have you been?" their teacher asked them. A few of them hedged; one said the city park; all of them began to drift away.

"It gets pretty competitive," Terry explained. "They won't even tell *me* where the best roosts are."

BACK IN THE CITY PARK at around ten, Callender's stu-dents kneeled on the grass comparing their hauls.

"I only got a hundred and fifty," said a skinny freshman boy wearing a Kansas State windbreaker.

"A hundred and ten," said a junior girl in a University of Kansas sweatshirt.

Terry shook his head. "Sorry we couldn't do better for you, ladies," he apologized to us. Then he spoke to his students: "OK, kids, how about handing over your envelopes to Dr. Perez?" They looked at him with a mixture of astonishment and betrayal. Hand over their butterflies? What was he thinking?

He tried again. "She needs them for an experiment she's doing. In Georgia."

Even so—give away their butterflies?

Reluctantly, they did.

THE EXPERIMENT almost worked. Perez sent collapsible tents and eighty butterflies packed on ice to a friend in the Washington, D.C., area, three days ahead of her own arrival. When she got there she was carrying an additional eighty monarchs. The next day she released them all and recorded which way they went. It wasn't a blind study—she knew which group was which, a factor that might have influenced the results—but the acclimatized ones did shift their orientation a little to the west, heading inland. It was a small shift, but statistically significant enough to warrant Perez's handing over a bottle of Dos Equis to Chip Taylor.

Georgia was a different matter altogether. Hurricane George was moving up the coast and Perez wasn't sure there would be enough time to acclimate the butterflies before the wind and rain began. So this time she brought both groups of monarchs with her, releasing the wild Kansas butterflies first, then waiting a few days before releasing the second set. The Kansas butterflies flew southwest. The others tended to the west, but only very slightly. It was inconclusive. When she did the distributions, Perez noticed something else, too. About half of the second group of butterflies were doing something totally un-predicted. They were heading south*east,* toward the ocean. To Sandra Perez this meant only one thing: she'd have to go back the next year and do the experiment again.

A DAY AWAY from Washington, D.C., as the monarch flies, in the Blue Ridge Mountains, Lincoln Brower was brooding. It was the end of the third week of September 1998, and no rain had fallen for nearly a month. He was standing at the Greenstone Overlook, at three thousand feet, looking down into the Shenandoah Valley. "Not one monarch," he said, shaking his head. "Nothing. Amazing." He got back into his car and continued on the Blue Ridge Parkway, then cut over to Skyline Drive, conducting an informal inventory. Loft Mountain, Big Run, Rockytop Ridge, Patterson Ridge—no monarchs. "Populations normally fluctuate tenfold," he said, driving down from Rocky Mountain. "If you had a hundred and fifty per man hour last year, fifteen would be low but within an accepted range. This year it's more like one point five."

Brower got out of the car and scanned the sky with his binoculars. In the three hours he had been driving around he had seen not a single monarch. "I don't know when it be-comes meaningful that it's a bad year," he said.

Chapter 10

BILL CALVERT EASED his truck off Interstate 281 near Johnson City, Texas, and headed out of town toward a spot on the map that appeared to be blank. Half an hour later he crossed over a cattle grate and under a sign that said Selah Ranch and proceeded along the scrub, past the live oaks, up to the main house. It was a few weeks shy of a year since our trip to Morelia. The same cassette tapes were sitting on the dashboard of his truck, waiting to be played. The same coverless, well-thumbed paperback *Random House Dictionary* was tucked under the seat. The same truck, nineteen thousand miles later. The nets, the roll of duct tape, the extension poles, the digital scale were all in the truckbed. Even the clothes Bill Calvert had on were the same. The changes had occurred elsewhere, away from the migratory part of his life, which remained constant.

It had been a bad year for monarchs, but not for Bill Calvert. He had his study site, his trips to Mexico; he had data to collate, papers to write. The ebb and flow of butterfly populations were background noise, at least so far. He had heard them before. It was enough to do the work.

Selah, where Calvert arrived at midafternoon, was a five-thousand-acre environmental education center in the Texas hill country. When Bill Calvert drove into the parking lot, he was greeted by Karen Oberhauser, Liz Goehring, and Sonia Altizer of the University of Minnesota. For the next six days they would be colleagues at what Calvert was calling Monarch Camp. It was Karen Oberhauser's show, really, part of her Minnesota-based Monarch Monitoring Project. The idea was to teach teachers and students together how to do science. Oberhauser, a Harvard graduate who'd been a schoolteacher before getting her doctorate, understood that for the most part teachers did not like to teach science. It scared them. The best way to demystify it, she thought, was to have them *do* it. The members of the group had already gotten together once, in Minnesota, and since then had been working on research projects. They would be continuing these at Selah and seeing the migration, too.

"In Minnesota we focused on breeding ecology and behavior," Oberhauser reminded the group. In shorts, sneakers, and a T-shirt, she looked the part of head counselor; the hand lens she was wearing on a lanyard around her neck could have been a whistle. "This week we're going to focus on migration, the nonbreeding part of the life cycle." She handed over the proceedings to Bill Calvert, introducing him as, among other things, the man who had found the Mexican overwintering sites. The young people, eager always to be touched by celebrity, sat up and took notice.

Calvert described a Texas flyway, three hundred miles wide from Wichita Falls to Eagle Pass, and talked about fall migration patterns and the spring remigration, about fire-ant predation, about how monarchs rode the updrafts along the spine of the Sierra Madre Oriental. The students listened politely, but what they wanted to know about more than anything was Calvert's initial discovery of the overwintering sites.

"The local people in Mexico thought the monarchs were coming there to die," he said. "They also thought the migrants were the souls of dead children. They usually began to arrive on November second, All Souls' Day." His audience laughed nervously. Butterflies were . . . butterflies. How could they be something—someone—else? But the students had been learning to do science using the null hypothesis, so they knew that the question could also be asked like this: How could the butterflies *not* be something—someone—else?

HOW DO PEOPLE KNOW what they know? This is always the question. The world presents itself: the sky is blue, the birds are singing. Our senses are an open window. A breeze is always blowing through. How do we know what cannot be proved? The answer is as unsatisfying as it is true: We just do. There are times when this is enough, and times when it is discomfiting. Recognition of a world that is not the familiar, material one is unsettling; it is hard enough to keep track of this world. Science, like belief, starts with wonder, and wonder starts with a question. As Bill Calvert would have told the students, answers did not dispel the wonder, they reinforced it. Answers begot questions, and questions were the libido of intelligence. How better to describe the endless pursuit of knowledge than passion?

•   •   •

THE QUESTION AT Monarch Camp the next day was how to test whether or not magnetism was a factor in monarch butterfly orientation. It was two-thirty in the afternoon, eighty-six degrees, with a slow wind boiling up out of the east. The students and teachers sat at picnic tables, pencils in hand as Liz Goehring and Sonia Altizer delivered a crash course in how to ask questions that could be answered by science.

"We're all brand new at studying orientation," Liz said, "so let's set up an experiment to figure out the effect of geomagnetism on the butterflies. What would be a good question to frame this investigation?"

" 'Will a strong magnetic pulse affect orientation?' " someone asked.

"Good," said Goehring. "That will work. We've got two kinds of monarchs—ones that have been raised in a greenhouse and wild ones—and we are going to expose only some of them to a magnetic pulse to confuse their polarity, so we'll have a control group. Can you think of a second question we should be looking at?"

" 'Will wild monarchs respond differently to being pulsed than ones raised in a greenhouse?' " a girl with a honeyed Texas accent asked.

Liz nodded her head yes. She wrote down the question. "So what do you expect will happen?"

The students raised their hands. "The wild ones that haven't been pulsed will fly to the southwest," one of them said.

"What about the other group?"

The participants weren't sure, so they settled on a random distribution: those butterflies might go anywhere.

"Let's find out," Liz said eagerly, directing everyone across the dirt road to an open field. Sonia was there already, surreptitiously exposing butterflies to a magnet so no one would know which monarchs were which. Launchpads (aka kitchen sponges) were distributed, and binoculars and compasses, too. Log sheets were drawn. The group arrayed itself across the field in pairs, and each pair was handed a butterfly in an envelope. They all studied their specimens and began recording information in their logs. Julia Goldberg, an eighth-grader from Rochester, Minnesota, took her first butterfly, knelt down behind it, and placed it on the sponge, which was facing east.

"Here goes," she said, letting loose its wings, which she had pinched together between her thumb and forefinger. She stepped back, expecting the monarch to rise up and make haste for the sky. But then a funny thing happened: the butterfly did not budge. It just sat on the sponge, casually—almost coquettishly—flapping its wings. A minute later it hopped off the sponge and pitched over into the grass. It looked drunk.

Julia released another butterfly and then another—eleven in all. Some took off immediately, heading south-southwest, a few lolled in the grass, and a couple more hung around the launchpad for a few minutes, then lifted off and aimed for the nearest tree. The same thing was happening all over the field. Maybe a dozen butterflies were scattered in the grass, and a bunch more were in the trees. As they sunned themselves, the data began to accumulate and a pattern began to emerge. It was like connect-the-dots before the last few numbers were reached: a picture was lurking there.

"Butterfly AA—what kind do you guess it is?" Liz asked

Julia, as the butterfly made no effort to move from its grassy perch.

"Wild, pulsed," she said confidently.

Liz checked her own data sheet and looked up in amazement. "Yes!" she said. "What about butterfly C?"

Julia consulted her log. "Lab-raised, no pulse," she ventured.

"Yes!" Liz said. "So what's the pattern?"

The eighth-grade science student looked over her notes. The pulsed butterflies didn't go anywhere, she explained; they just hung out in the grass. The unpulsed lab-raised monarchs didn't go far at first either, but then, after warming themselves in the sun for a few minutes and shaking off the chill of the cooler in which they'd been kept, they flew off in no particular direction. In contrast, the unpulsed wild butterflies took off right away, heading southwest. Liz was excited. Julia was excited. Her guesses were right. She was as giddy as a stock picker whose system was working. If she was ever going to get hooked on science, this was the moment.

A look of concern, a small one, crossed Liz's face. "It doesn't always work out this neatly," she cautioned. Julia and the other members of the group nodded responsibly. But then Liz was smiling again. The fact was, it *had* worked out. Who could say it wouldn't again? The kids and teachers gave one another high-fives.

NEAT AS THE experiment had been, a problem began to nag at me. By then the students were on to other things, though the glow of success still hung over them. But the question for me was, what did it mean that butterflies exposed to a magnetic pulse fell headlong into the grass? Not, certainly, that monarch butterflies used magnetism to orient themselves in flight. No experiment had yet been designed to prove that. All the experiment seemed to demonstrate was that monarchs had magnetite in their bodies that responded to a magnetic charge. But that was already well known. Interfering with a function did not prove that that function, when intact, was essential. Drinking alcohol would impair my ability to drive, for instance, but *not* drinking alcohol was not what enabled me to drive carefully (or not) at other times.

That night I put my question to Bill Calvert, who was patiently gluing dead monarchs onto a strip of cardboard. "You're probably right," he said, shrugging. "But it was pretty great, wasn't it?"

Knowledge can be passed along from person to person like a baton in a race, but the pursuit of knowledge, and love for the pursuit of knowledge, that particular passion, can only be chosen. This, I think, was what Bill Calvert was hinting at, or what I wanted him to mean.

THERE WERE MONARCHS migrating through at Selah, but not many, not enough, and not in the kinds of numbers that might make an impression. So the campers were loaded into vans and Bill Calvert got into his truck. He knew a place a few hours away, a reliable roost at Garner State Park. It would be an overnight field trip.

Sandra Perez had arrived that afternoon on her way to a meeting farther south, so she came along, too. Her Georgia experiment was finished, the monarch season was nearly over, and neither had gone exactly the way anyone might have predicted. Soon Perez would have to get on with her real work, her postdoc on ants. If there was going to be a valedictory to this elusive generation of monarchs, this would be the last chance.

It was a long drive into the heart of the hill country, through Twin Sisters and Comfort, along the Guadalupe River and under the junipers in its fertile valley. It was nearly dark when the group arrived at Garner State Park, and Calvert hustled everyone into the woods. In the quickening night, students and teachers gathered around him—first a ring of students, then a ring of teachers. The question of the moment had to do with roost formation, Bill explained. Monarchs roosted in groups. Were they attracted to big clusters, to small clusters, or to no clusters at all? He held up one of the strips of dead monarchs he had been assembling the night before. A dozen open-winged butterflies were crowded together on it, back to front, like moviegoers queued on opening night.

"We'll use these as decoys," Bill said, showing how to attach a strip to a tree branch so it hung along the bias in the right direction. "We're going to observe whether, when the monarchs see the clusters, they'll try to join the roost."

For the tree climbers among the group it was an opportune time to show off. Up they went, shimmying out on the limbs of live oaks to fasten the decoys onto the leaves. For those who couldn't climb, the decoys were placed at arm's height on pecan trees—a design flaw, no doubt, since monarchs were not likely to roost so low. But that was beside the point. Fifteen minutes passed, then twenty-five. The occasional monarch fell through the canopy and flew about, but most of those that did just left again, and none made even a wink at the decoy clusters. The experiment was canceled on account of the fact that the visiting team did not show up. Nor did the home team, for that matter.

"From the reports I was getting, I would have guessed that there were gobs and gobs of monarchs here recently," Bill Calvert said apologetically. "We must be in the trough of the wave. It's too bad, but it's nature."

"LET'S GET OUT of here and go to Del Rio," I said to Sandra Perez then, though it was after nine and Del Rio, the Texas town on the Mexican border through which millions of monarchs regularly passed on their way to the other side, was at least four hours away. As a child I had often read the last page of a book when I was in the thick of it, and even when I hadn't, that last page had still been there to be resisted or given in to. Del Rio was the final page of this year's volume on monarchs, and though it wouldn't conclude anything, really, I suddenly needed to get there.

We drove through the night trying to outdistance the butterflies, somewhere acquiring a bottle of scotch and a block of cheese and a cheap motel room containing a mismatched set of dinner plates. I had seen butterflies before. I had seen gobs and gobs of butterflies before, but for me this trip to the border was like skipping to the end of the book. It had been months since the monarchs had left Mexico, and I just wanted to know one thing: Were they here? Were the butterflies I hadn't seen in the Adirondacks and at Cape May and in Toronto and Lawrence up in the sky? It wouldn't *mean* anything if they were or if they weren't, not in a scientific way, but it would mean something to me. I had been watching monarchs for years. Their story was part of mine, and I wanted to *know*, as if knowledge were a promise, not a set of facts.

•   •   •

IN THE MORNING we continued on for another forty-five miles, to Seminole Canyon, a limestone basin out on Route 90. It was raining lightly, and the sky was overcast, and the wind was blowing from the southeast—not ideal conditions for a monarch butterfly.

"Last year there were millions here in the tree by the ranger's station," a man in a camper told us when we asked if he'd seen any butterflies. "This year they've just been filtering through the canyon, out of the wind." He directed us to get back in the car and drive down to where the Pecos River met the Rio Grande.

The rain was played out. We drove in silence, down a steep hill to a dry riverbed where sheer rock walls rose up from the ground like battlements. It was quiet down there. The grasses had grown high but couldn't quite cover the rusted cans and yellowed plastic bottles that littered the ground. Nature had reasserted itself, but not enough. We stepped over an abandoned boat launch and followed a path down to the bottom. There was no wind. Not a lick. No one was around. It was better that way.

We climbed back up and sat on a stone jetty. I handed my binoculars to Sandra, who leaned back and lifted them to her eyes. The last page was upon us, and I wanted her to tell me what was written there. Without a word she handed the glasses back to me. The sky was steel gray. Mexico was a short swim away. Air was a fluid. I took a look, and then another, and settled in to look some more. There they were, the monarchs, a steady stream of them, twenty-three, forty-four, one hundred and nine, and then I lost count.

NOTE ON RESOURCES

Of the many books and articles I have read about monarch butterfly migration, the first, Fred Urquhart's *The Monarch Butterfly: International Traveler* (William Caxton, Ltd.), remains one of the most interesting from a historical perspective, in chronicling the Urquharts' early efforts to understand monarch migration. A more comprehensive history of the quest to track these butterflies, as well as an introduction to the naturalists whose quest it has been, may be found in Lincoln P. Brower's definitive essay "Understanding and Misunderstanding the Migration of the Monarch Butterfly in North America, 1857–1994," in the *Journal of the Lepidopterists' Society*, no. 49 (1995). Although they are too numerous to list here, Brower's scientific writings, which touch on almost every aspect of monarch biology and behavior, are essential reading for anyone hoping to gain more than a superficial understanding of *Danaus plexippus*. For example, his early collaborations with Jane Brower, published in *Ecology*, no. 43 (1962), and the *Journal of the Lepidopterists' Society*, no. 15 (1961), among others, provide useful background on mimicry and are resonant with his later articles written with

Linda Fink (*Nature,* no. 291 [1981]) and J. N. Seiber et al. (*Journal of Chemical Ecology,* nos. 10 [1984] and 8 [1981]), on cardenolides. Lincoln Brower's work with William Calvert (in the *Journal of the Lepidopterists' Society,* nos. 46 [1992], 43 [1989], and 40 [1986]; *Biotropica,* no. 15 [1983]; and *Science,* no. 204 [1979], to name just a few) opens wide a window on the Mexican overwintering sites and the dangers faced by monarch butterflies there.

The two large and exhaustive volumes on biology, sociology, and conservation that have come out of the three international symposia on monarch butterflies held in 1981, 1986, and 1997 are crucial from both scientific and historical standpoints. These are *Biology and Conservation of the Monarch Butterfly,* edited by Stephen Malcolm and Myron Zalucki (Science Series no. 38, Natural History Museum of Los Angeles County), and the *Proceedings of the 1997 North American Conference on the Monarch Butterfly,* edited by Jurgen Hoth, Leticia Merino, Karen Oberhauser et al. (Commission for Environmental Cooperation, Montreal, Canada). A more accessible (and beautifully illustrated) introduction to monarchs is Eric Grace's *The World of the Monarch Butterfly* (Sierra Club Books); also recommended is Kathryn Lasky's children's book *Monarchs* (Gulliver Books).

In the midst of all this prose, I have taken solace in Alison Hawthorne Deming's *The Monarchs: A Poem Sequence* (Louisiana State University Press) and in the work of the esteemed Mexican poet Homero Aridjis, whose collection *Exaltation of Light* is available in English. While the poetry of Aridjis's good friend W. S. Merwin provided many wonderful literary diversions, his essay on monarch butterflies in *Orion* magazine (Winter 1996) was critical to this project.

In many ways my interest in monarchs was piqued by a number of magnificent books about birds. These included Kenn Kaufman's *Kingbird Highway: The Story of a Natural Obsession That Got Out of Hand* (Houghton Mifflin), Jonathan Weiner's *The Beak of the Finch: A Story of Evolution in Our Time* (Vintage Books), Christopher Cokinos's *Hope Is the Thing with Feathers: A Personal Chronicle of Vanishing Birds* (Jeremy Tarcher), Paul Kerlinger's *How Birds Migrate* (Stackpole Books), and Scott Weidensaul's *Living on the Wind: Across the Hemisphere with Migratory Birds* (North Point Press). Werner Nachtigall's entomological study *Insects in Flight* also nudged my increasing preoccupation with both aviation and butterflies, as did the work of Gary Paul Nabhan, especially *The Forgotten Pollinators* (with Stephen Buchmann, Island Press) and Sue Hubbell's *Broadsides from the Other Orders: A Book of Bugs* (Houghton Mifflin).

Three field guides proved invaluable: *The Audubon Society Handbook for Butterfly Watchers* (Scribner's), by Robert Michael Pyle; the same author's *Audubon Field Guide to North American Butterflies* (Knopf); and Jeffrey Glassberg's *Butterflies through Binoculars* (Oxford). The work of Professor Adrian Wenner on monarchs, honeybees, and epistemology offered a necessary and provocative challenge to the "truth" of monarch research in particular and of science in general. Particularly helpful were his book *Anatomy of a Controversy: The Question of a "Language" among Bees* (with Patrick Wells; Columbia University Press) and a number of as-yet-unpublished responses to the orientation experiments by Sandra Perez and Chip Taylor described in their paper "The Sun Compass in Monarch Butterflies" (*Nature,* no. 387, [1997]). Wenner's studies of California monarchs supplied a useful perspective on the West Coast population, as did *Monarch News,* a monthly newsletter published by California Monarch Studies, a not-for-profit organization run by the avid and prolific San Diego lepidopterist David Marriott. By far the most intimate and insightful understanding of western monarch behavior, and of the quaint and obsessive behavior of one of our great entomologists, is to be found in Robert Michael Pyle's personal epic *Chasing Monarchs: A Migration with the Butterflies of Passage* (Houghton Mifflin). Even farther west, University of Hawaii professor John Stimson's pioneering research on white monarchs introduced me to genetic mutation, polymorphism, and the role of predation in biological diversity.

Resources available on the Internet have been essential to this project and will, I wager, play an increasing role in the collection and dissemination of knowledge about monarchs. The University of Kansas's Monarch Watch Web site (www.monarchwatch.org) and its e-mail discussion group, D-Plex, as well as the organization's newsletter, are the crossroads for almost all monarch enthusiasts, whether professional or amateur. Monarch Watch, along with Journey North (www.learner.org/jnorth), not only track the movement of monarchs but also archive vital scientific papers and historical data. Both sites were of inestimable value to me.

I also relied on the systematic longitudinal data on the eastern coastal migration collected by Dick Walton's Monarch Monitoring Project, formerly called the Monarch Migration Association of North America (www.concord.org/~dick/mon.html), and on the comprehensive account of flying and gliding, wind and weather, provided on David Gibo's "Tactics and Vectors" Web site (www.erin.utoronto.ca/~w3gibo), which also shares data on monarch movements and corollary meteorological conditions relayed by pilots. For information on the effect of genetically altered corn pollen on monarch larvae, I consulted the Website run by the University of Iowa's entomology department (www.ent.iastate.edu). The eponymous "butterflywebsite" (www.butterflywebsite.com), which is run by commercial breeders, not only offered insight into the controversial world of butterfly farming, it supplied an extensive list of butterfly houses around the world, online reference material about butterflies and moths, and links to international entomological organizations. Finally, the University of Minnesota's Monarch Lab (www.monarchlab.umn.edu) provided curricular information, archival data, experimental data, and, perhaps most important, links to the ever-expanding universe of monarch biology.

ACKNOWLEDGMENTS

When I blithely followed Lincoln Brower up that mountain in Mexico years ago, I had no idea I had come to a path I would still be traversing to this day. For this gift, especially, I am thankful, and for his scholarship and intellectual ardor, which he has generously shared with me over the years. Many others in the monarch community, both professional and amateur lepidopterists, were instrumental in trying to keep me on track, particularly Chip Taylor and the staff of Monarch Watch, Sandra Perez, Karen Oberhauser, Bob Pyle, Homero and Betty Aridjis, William Merwin, Monica Missrie, Richard Walton, Elizabeth Howard at Journey North, Don Davis, Paul Cherubini, Dan Petr, Dale Clayton, Ro Vaccaro, Steve Montgomery, John Stimson, David Marriott, Adrian Wenner, David Gibo, the late Ken Brugger, David Barkin, and Terry Callender and his students at Wamego High School. Bill Calvert stands alone. I am grateful that he let me go along for the ride, and for his wisdom, kindness, and (two sigmas above a seventeen-year-old's) daring.

I am indebted, as well, to the John Simon Guggenheim Foundation, whose fellowship support was essential to this endeavor, and to the echoing green and Blessing Way foundations, which found me at critical moments.

My editorial confreres over the years at *DoubleTake,* Robert Coles, Alex Harris, Rob Odom, and David Rowell, gave me the time and space to explore my growing preoccupation with monarch butterflies, as did Roger Cohn, who, first at *Audubon* and then at *Mother Jones,* supported this project from its inception, and Irene Schneider and Tom Wallace at *Condé Nast Traveler,* who flung me far afield.

Thanks, always, to Dan Frank at Pantheon, who got it, stayed with it, and made it better, and to my patient, farsighted friend Gloria Loomis and her assistant Katherine Faussett at the Watkins-Loomis Agency. Thanks, too, to Rebecca Wilson at Weidenfeld and Nicolson in England, and to Abner Stein, who put me in her care.

The unfailing encouragements of Sara Rimer and Nicky Dawidoff were inestimable, as were the guidance and help of Barbara Epstein, John Elder, Francine Prose, Edward Hirsch, Harriet Barlow, Jane Mayer, Tony Horwitz, Geraldine Brooks, Terry and Brooke Williams, Shawn Leary, Michael Considine, Sam and Lisa Verhovek, David Goldfarb, Lisa Saiman, Kathy and Gary Wilson, Joan Reynolds, Russell Puschak, Kate Gardner, Jackie and Nick Avignon, Joanne Rizzi, Diane Willow, Sally Jacobs, Ellen Fitzpatrick, Michael Dabroski, Lisa Spilde, Liz Phillips, Susan Mintz, Bill Jaeger, Amanda Smith-Socaris, Linda Motzkin, Jonathan Rubenstein, Nancy Dreyer, Gabriel Dreyer, and Joan Koffman.

My fellow trustees of the Town of Johnsburg Library, as well as the library's remarkable staff, kept me in books and good cheer, while Jenna Stauffer and the students of the Wilderness Community School in Johnsburg, New York, and Robin Gunn and Jill Ferraresso's Y2K Explorers' Class at the Atrium School in Watertown, Massachusetts—Ben Adams-Keane, Ali Broadstone, Esther Brown, Axelle Derviscevic, Tamara Ivanovic, Briana Karman, Sophie McKibben, Miles Powell, Ivy Olcott, Lily Moore, Etta Resnick-Field, Max Newman, Brennan Robbins, Michael Slonina, Lucien Swetshinski, and Emma Wulf—kept me on my toes.

Thanks beyond words to Bill for his love, care, and constancy, and to his parents and mine for theirs. This book is dedicated to two keen observers of natural phenomena, Sophie and Barley, who have shown me so much.